Deciphering Mineralogy

Edited by **John Wayne**

New York

Published by Callisto Reference,
106 Park Avenue, Suite 200,
New York, NY 10016, USA
www.callistoreference.com

Deciphering Mineralogy
Edited by John Wayne

International Standard Book Number: 978-1-63239-145-2 (Hardback)

Printed in the United States of America.

Contents

Preface

This book is a compilation of easy to comprehend researches. It offers an extensive account of the field of mineralogy. This book targets chemists, physicists, engineers and students of geology, geophysics and soil science. It is vital to the advanced students of mineralogy who seek concise yet precise content. They can refer to it to broaden their knowledge or as a reference in their work.

Significant researches are present in this book. Intensive efforts have been employed by authors to make this book an outstanding discourse. This book contains the enlightening chapters which have been written on the basis of significant researches done by the experts.

Finally, I would also like to thank all the members involved in this book for being a team and meeting all the deadlines for the submission of their respective works. I would also like to thank my friends and family for being supportive in my efforts.

Editor

An Introduction to Mineralogy

Cumhur Aydinalp

Uludag University, Bursa,
Turkey

1. Introduction

The science of mineralogy is a branch of the earth sciences that is concerned with studying minerals and their physical and chemical properties. Within mineralogy there are also those who study how minerals are formed, where they are geographically located, as well as their potential uses. Like many sciences, mineralogy has its origins in several ancient civilizations, and it has been concerned primarily with the various methods of classification of minerals for most of its history. Modern-day mineralogy has been expanded by advances in other sciences, such as biology and chemistry, to shed even more light on the nature of the materials that form the earth we live on.

The ancient Greek philosopher Aristotle was one of the first people to theorize extensively about the origins and properties of minerals. His ideas were new and advanced for the time, but he and his contemporaries were largely incorrect in their assumptions. For example, it was a widely held belief in ancient Greece that the mineral asbestos was a kind of vegetable. Nevertheless, these ancient theories provided a starting point for the evolution of mineralogy as we have come to know it. It was not until the 16th century that mineralogy began to take a form that is recognizable to us, largely thanks to the work of German scientist Georgius Agricola.

- For example, it was a widely held belief in ancient Greece that the mineral asbestos was a kind of vegetable. Nevertheless, these ancient theories provided a starting point for the evolution of mineralogy as we have come to know it. It was not until the 16th century that mineralogy began to take a form that is recognizable to us, largely thanks to the work of German scientist Georgius Agricola.

2. Definition of mineral

A mineral is a naturally-occurring, homogeneous solid with a definite, but generally not fixed, chemical composition and an ordered atomic arrangement. It is usually formed by inorganic processes.

Let's look at the five parts of this definition:

1. "Naturally occurring" means that synthetic compounds not known to occur in nature cannot have a mineral name. However, it may occur anywhere, other planets, deep in the earth, as long as there exists a natural sample to describe.

2. "Homogeneous solid" means that it must be chemically and physically homogeneous down to the basic repeat unit of the atoms. It will then have absolutely predictable physical properties (density, compressibility, index of refraction, etc.). This means that rocks such as granite or basalt are not minerals because they contain more than one compound.

3. "Definite, but generally not fixed, composition" means that atoms, or groups of atoms must occur in specific ratios. For ionic crystals (i.e. most minerals) ratios of cations to anions will be constrained by charge balance, however, atoms of similar charge and ionic radius may substitute freely for one another; hence definite, but not fixed.

4. "Ordered atomic arrangement" means crystalline. Crystalline materials are three-dimensional periodic arrays of precise geometric arrangement of atoms. Glasses such as obsidian, which are disordered solids, liquids (e.g., water, mercury), and gases (e.g., air) are not minerals.

5. "Inorganic processes" means that crystalline organic compounds formed by organisms are generally not considered minerals. However, carbonate shells are minerals because they are identical to compounds formed by purely inorganic processes.

An abbreviated definition of a mineral would be "a natural, crystalline phase". Chemists have a precise definition of a phase. A phase is that part of a system which is physically and chemically homogeneous within itself and is surrounded by a boundary such that it is mechanically separable from the rest of the system. The third part of our definition of a mineral leads us to a brief discussion of stoichiometry, the ratios in which different elements (atoms) occur in minerals. Because minerals are crystals, dissimilar elements must occur in fixed ratios to one another. However, complete free substitution of very similar elements (e.g., Mg^{+2} and Fe^{+2} which are very similar in charge (valence) and radius is very common and usually results in a crystalline solution (solid solution). For example, the minerals forsterite (Mg_2SiO_4) and fayalite (Fe_2SiO_4) are members of the olivine group and have the same crystal structure, that is, the same geometric arrangement of atoms. Mg and Fe substitute freely for each other in this structure, and all compositions between the two extremes, forsterite and fayalite, may occur. However, Mg or Fe do not substitute for Si or O, so that the three components, Mg/Fe, Si and O always maintain the same 2 to 1 to 4 ratio because the ratio is fixed by the crystalline structure. These two minerals are called end-members of the olivine series and represent extremes or "pure" compositions. Because these two minerals have the same structure, they are called isomorphs and the series, an isomorphous series.

In contrast to the isomorphous series, it is also common for a single compound (composition) to occur with different crystal structures. Each of these structures is then a different mineral and, in general, will be stable under different conditions of temperature and pressure. Different structural modifications of the same compound are called polymorphs. An example of polymorphism is the different minerals of SiO_2 (silica); alpha-quartz, beta-quartz, tridymite, cristobalite, coesite, and stishovite. Although each of these has the same formula and composition, they are different minerals because they have different crystal structures. Each is stable under a different set of temperature and pressure conditions, and the presence of one of these in a rock may be used to infer the conditions of formation of a rock. Another familiar example of polymorphism is graphite and diamond, two different minerals with the same formula, C (carbon).

Glasses (obsidian), liquids, and gases however, are not crystalline, and the elements in them may occur in any ratios, so they are not minerals. So in order for a natural compound to be a mineral, it must have a unique composition and structure (Blackburn & Dennen, 1988).

3. Composition of the earth's crust

The earth's crust is composed of many kinds of rocks, each of which is an aggregate of one or more minerals. In geology, the term mineral describes any naturally-occurring solid substance with a specific composition and crystal structure. A mineral's composition refers to the kinds and proportions of elements making up the mineral. The way these elements are packed together determines the structure of the mineral. More than 3,500 different minerals have been identified. There are only 12 common elements (oxygen, silicon, aluminum, iron, calcium, sodium, potassium, magnesium, titanium, hydrogen, manganese, phosphorus) that occur in the earth's crust. All other naturally occurring elements are found in very minor or trace amounts. Silicon and oxygen are the most abundant crustal elements, together comprising more than 70 percent by weight (Rudnick & Fountain, 1995). It is therefore not surprising that the most abundant crustal minerals are the silicates (e.g. olivine, Mg_2SiO_4), followed by the oxides (e.g. hematite, Fe_2O_3). Other important types of minerals include: the carbonates (e.g. calcite, $CaCO_3$) the sulfides (e.g. galena, PbS) and the sulfates (e.g. anhydrite, $CaSO_4$). Most of the abundant minerals in the earth's crust are not of commercial value. Economically valuable minerals (metallic and nonmetallic) that provide the raw materials for industry tend to be rare and hard to find. Therefore, considerable effort and skill is necessary for finding where they occur and extracting them in sufficient quantities. Table 1 shows the elemental chemical composition of the Earth's crust in order of abundance (Lutgens & Tarbuck, 2000).

Element name	Symbol	Percentage by weight of the Earth's crust
Oxygen	O	46,6
Silicon	Si	27,7
Aluminium	Al	8,1
Iron	Fe	5,0
Calcium	Ca	3.6
Sodium	Na	2,8
Potassium	K	2,6
Magnesium	Mg	2,1
All other elements		1,5

Table 1. The elements in the Earth's crust (Lutgens & Tarbuck, 2000).

This is a table that shows the elemental chemical composition of the Earth's crust. They will vary depending on the way they were calculated and the source. 98.5% of the Earth's crust consists of oxygen, silicon, aluminum, iron, calcium, sodium, potassium and magnesium. All other elements account for approximately 1.5% of the volume of the Earth's crust.

4. The some characteristics of minerals

The physical properties of a mineral are determined by its chemical composition and its crystalline structure. Within the limits of the permissible variation in chemical composition,

different samples of a single mineral species are expected to display the same set of physical properties. These characteristic physical properties are therefore very useful to the field geologist in identifying and describing a specimen (Zoltai & Stout,1984).

Properties which describe the physical appearance of a mineral specimen include color, streak, and luster. Mass-dependent properties include density; mechanical properties include hardness, cleavage, fracture, and tenacity. Properties relating to the growth patterns and physical appearance of crystals, both individually and in aggregate, are described in terms of crystal habit, crystal form, and crystal system (Klein & Hurlbut, 1985).

1. Crystal form and habit (shape).
2. Luster and transparency
3. Color and streak.
4. Cleavage, fracture, and parting.
5. Tenacity
6. Density
7. Hardness

4.1 Crystal form and habit

The crystal faces developed on a specimen may arise either as a result of growth or of cleavage. In either case, they reflect the internal symmetry of the crystal structure that makes the mineral unique. The crystal faces commonly seen on quartz are growth faces and represent the slow est growing directions in the structure. Quartz grows rapidly along its c-axis (three-fold or trigonal symmetry axis) direction and so never shows faces perpendicular to this direction. On the other hand, calcite rhomb faces and mica plates are cleavages and represent the weakest chemical bonds in the structure. There is a complex terminology for crystal faces, but some obvious names for faces are prisms and pyramids. A prism is a face that is perpendicular to a major axis of the crystal, whereas a pyramid is one that is not perpendicular to any major axis.

Crystals that commonly develop prism faces are said to have a prismatic or columnar habit. Crystals that grow in fine needles are acicular; crystals growing flat plates are tabular. Crystals forming radiating sprays of needles or fibers are stellate. Crystals forming parallel fibers are fibrous, and crystals forming branching, tree-like growths are dendritic.

4.2 Luster and transparency

The way a mineral transmits or reflects light is a diagnostic property. The transparency may be either opaque, translucent, or transparent. This reflectance property is called luster. Native metals and many sulfides are opaque and reflect most of the light hitting their surfaces and have a metallic luster. Other opaque or nearly opaque oxides may appear dull, or resinous. Transparent minerals with a high index of refraction such as diamond appear brilliant and are said to have an adamantine luster, whereas those with a lower index of refraction such as quartz or calcite appear glassy and are said to have a vitreous luster.

4.3 Color and streak

Color is fairly self-explanatory property describing the reflectance. Metallic minerals are either white, gray, or yellow. The presence of transition metals with unfilled electron shells

(e.g. V, Cr, Mn, Fe, Co, Ni, and Cu) in oxide and silicate minerals causes them to be opaque or strongly colored so that the streak, the mark that they leave when scratched on a white ceramic tile, will also be strongly colored.

4.4 Cleavage, fracture, and parting

Because bonding is not of equal strength in all directions in most crystals, they will tend to break along crystallographic directions giving them a fracture property that reflects the underlying structure and is frequently diagnostic. A perfect cleavage results in regular flat faces resembling growth faces such as in mica, or calcite. A less well developed cleavage is said to be imperfect, or if very weak, a parting. If a fracture is irregular and results in a rough surface, it is hackly. If the irregular fracture propagates as a single surface resulting in a shiny surface as in glass, the fracture is said to be conchoidal.

4.5 Tenacity

Tenacity is the ability of a mineral to deform plastically under stress. Minerals may be brittle, that is, they do not deform, but rather fracture, under stress as do most silicates and oxides. They may be sectile, or be able to deform so that they can be cut with a knife. Or, they may be ductile and deform readily under stress as does gold.

4.6 Density

Density is a well-defined physical property measured in g/cm^3. Most silicates of light element have densities in the range 2.6 to 3.5. Sulfides are typically 5 to 6. Iron metal about 8, lead about 13, gold about 19, and osmium, the densest substance, and a native element mineral is 22.

4.7 Hardness

Hardness is usually tested by seeing if some standard minerals are able to scratch others. A standard scale was developed by Friedrich Mohs in 1812. The standard minerals making up the Mohs scale of hardness are:

1. Talc
2. Gypsum
3. Calcite
4. Fluorite
5. Apatite

6. Orthoclase
7. Quartz
8. Topaz
9. Corundum
10 Diamond

This scale is approximately linear up to corundum, but diamond is approximately 5 times harder than corundum.

4.8 Unique properties

A few minerals may have easily tested unique properties that may greatly aid identification. For example, halite (NaCl) (common table salt) and sylvite (KCl) are very similar in most of their physical properties, but have a distinctly different taste on the tongue, with sylvite having a more bitter taste. Another unique property that can be used to distinguish between

otherwise similar back opaque minerals is magnetism. For example, magnetite (Fe_3O_4), ilmenite ($FeTiO_3$), and pyrolusite (MnO_2) are all dense, black, opaque minerals which can easily be distinguished by testing the magnetism with a magnet. Magnetite is strongly magnetic and can be permanently magnetized to form a lodestone; ilmenite is weakly magnetic; and pyrolusite is not magnetic at all.

4.9 Other properties

There are numerous other properties that are diagnostic of minerals, but which generally require more sophisticated devices to measure or detect. For example, minerals containing the elements U or Th are radioactive, and this radioactivity can be easily detected with a Geiger counter. Examples of radioactive minerals are uraninite (UO_2), thorite ($ThSiO_4$), and carnotite ($K_2(UO_2)(VO_4)_2$ rH_2O). Some minerals may also be fluorescent under ultraviolet light, that is they absorb UV lighta and emit in the visible. Other optical properties such as index of refraction and pleochroism (differential light absorption) require an optical microscope to measure. Electrical conductivity is an important physical property but requires an impedance bridge to measure. In general native metals are good conductors, sulfides of transition metals are semi-conductors, whereas most oxygen-bearing minerals (i.e., silicates, carbonates, oxides, etc.) are insulators. Additionally, quartz (SiO_2) is piezoelectric (develops an electrical charge at opposite end under an applied mechanical stress); and tourmaline is pyroelectric (develops an electrical charge at opposite end under an applied thermal gradient).

5. Mineral occurences and environments

In addition to physical properties, one of the most diagnostic features of a mineral is the geological environment in which it is occurs (Deer, Howie & Zussman, 1992).

5.1 Igneous minerals

Minerals in igneous rocks must have high melting points and be able to co-exist with, or crystallize from, silicate melts at temperatures above 800 ° C. Igneous rocks can be generally classed according to their silica content with low-silica (< 50 % SiO_2) igneous rocks being termed basic or mafic, and high-silica igneous rocks being termed **silicic** or **acidic**. Basic igneous rocks (BIR) include basalts, dolerites, gabbros, kimberlites, and peridotites, and abundant minerals in such rocks include olivine, pyroxenes, Ca-feldspar (plagioclase), amphiboles, and biotite. The abundance of Fe in these rocks causes them to be dark-colored. Silicic igneous rocks (SIR) include granites, granodiorites, and rhyolites, and abundant minerals include quartz, muscovite, and alkali feldspars. These are commonly light-colored although color is not always diagnostic. In addition to basic and silicic igneous rocks, a third igneous mineral environment representing the final stages of igneous fractionation is called a pegmatite (PEG) which is typically very coarse-grained and similar in composition to silicic igneous rocks (i.e. high in silica). Elements that do not readily substitute into the abundant minerals are called incompatible elements, and these typically accumulate to form their own minerals in pegmatites. Minerals containing the incompatible elements, Li, Be, B, P, Rb, Sr, Y, Nb, rare earths, Cs, and Ta are typical and characteristic of pegmatites.

5.2 Metamorphic minerals

Minerals in metamorphic rocks have crystallized from other minerals rather than from melts and need not be stable to such high temperatures as igneous minerals. In a very general way, metamorphic environments may be classified as low-grade metamorphic (LGM) (temperatures of 60 ° to 400 ° C and pressures < .5 GPa (=15km depth) and high-grade meta morphic (HGM) (temperatures > 400 ° and/or pressures > .5 GPa). Minerals characteristic of low- grade metamorphic environments include the zeolites, chlorites, and andalusite. Minerals characteristic of high grade metamorphic environments include sillimanite, kyanite, staurolite, epidote, and amphiboles.

5.3 Sedimentary minerals

Minerals in sedimentary rocks are either stable in low-temperature hydrous environments (e.g. clays) or are high temperature minerals that are extremely resistant to chemical weathering (e.g. quartz). One can think of sedimentary minerals as exhibiting a range of solubilities so that the most insoluble minerals such as quartz, gold, and diamond accumulate in the coarsest detrital sedimentary rocks, less resistant minerals such as feldspars, which weather to clays, accumulate in finer grained siltstones and mudstones, and the most soluble minerals such as calcite and halite (rock-salt) are chemically precipitated in evaporite deposits. Sedimentary minerals can classify into detrital sediments (DSD) and evaporites (EVP). Detrital sedimentary minerals include quartz, gold, diamond, apatite and other phosphates, calcite, and clays. Evaporite sedimentary minerals include calcite, gypsum, anhydrite, halite and sylvite, plus some of the borate minerals.

5.4 Hydrothermal minerals

The fourth major mineral environment is hydrothermal, minerals precipitated from hot aqueous solutions associated with emplacement of intrusive igneous rocks. This environment is commonly grouped with metamorphic environments, but the minerals that form by this process and the elements that they contain are so distinct from contact or regional metamorphic rocks that it us useful to consider them as a separate group. These may be sub-classified as high temperature hydrothermal (HTH), low temperature hydrothermal (LTH), and oxydized hydrothermal (OXH). Sulfides may occur in igneous and metamorphic rocks, but are most typically hydrothermal. High temperature hydrothermal minerals include gold, silver, tungstate minerals, chalcopyrite, bornite, the tellurides, and molybdenite. Low temperature hydrothermal minerals include barite, gold, cinnabar, pyrite, and cassiterite. Sulfide minerals are not stable in atmospheric oxygen and will weather by oxidation to form oxides, sulfates and carbonates of the chalcophile metals, and these minerals are characteristic of oxidized hydrothermal deposits. Such deposits are called gossans and are marked by yellow-red iron oxide stains on rock surfaces. These usually mark mineralized zones at depth.

6. The mineral classification

Minerals are classified on their chemistry, particularly on the anionic element or polyanionic group of elements that occur in the mineral. An anion is a negatively charge atom, and a

polyanion is a strongly bound group of atoms consisting of a cation plus several anions (typically oxygen) that has a net negative charge.

For example carbonate $(CO_3)^{2-}$, silicate $(SiO_4)^{4-}$ are common polyanions. This classification has been successful because minerals rarely contain more than one anion or polyanion, whereas they typically contain several different cations (Nesse, 2000).

6.1 Native elements

The first group of minerals is the native elements, and as pure elements, these minerals contain no anion or polyanion. Native elements such as gold (Au), silver (Ag), copper (Cu), and platinum (Pt) are metals, graphite is a semi-metal, and diamond (C) is an insulator.

6.2 Sulfides

The sulfides contain sulfur (S) as the major "anion". Although sulfides should not be considered ionic, the sulfide minerals rarely contain oxygen, so these minerals form a chemically distinct group. Examples are pyrite (FeS_2), sphalerite (ZnS), and galena (PbS). Minerals containing the elements As, Se, and Te as "anions" are also included in this group.

6.3 Halides

The halides contain the halogen elements (F, Cl, Br, and I) as the dominant anion. These minerals are ionically bonded and typically contain cations of alkali and alkaline earth ele ments (Na, K, and Ca). Familiar examples are halite (NaCl) (rock salt) and fluorite (CaF_2).

6.4 Oxides

The oxide minerals contain various cations (not associated with a polyanion) and oxygen. Examples are hematite (Fe_2O_3) and magnetite (Fe_3O_4).

6.5 Hydroxides

These minerals contain the polyanion OH- as the dominant anionic species. Examples include brucite $(Mg(OH)_2)$ and gibbsite $(Al(OH)_3)$.

6.6 Carbonates

The carbonates contain CO_3^{2-} as the dominant polyanion in which C^{4+} is surrounded by three O^{2-} anions in a planar triangular arrangement. A familiar example is calcite $(CaCO_3)$. Because NO_3^- shares this geometry, the nitrate minerals such as soda niter (nitratite) $(NaNO_3)$ are included in this group.

6.7 Sulfates

These minerals contain SO_4^{2-} as the major polyanion in which S^{6+} is surrounded by four oxygen atoms in a tetrahedron. Note that this group is distinct from sulfides which contain no O. A familiar example is gypsum $(CaSO_4.2H_2O)$.

6.8 Phosphates

The phosphates contain tetrahedral PO_4^{3-} groups as the dominant polyanion. A common example is apatite $(Ca_5(PO_4)_3(OH))$ a principal component of bones and teeth. The other trivalent tetrahedral polyanions, arsenate AsO_4^{3-}, and vanadate VO_4^{3-} are structurally and chemically similar and are included in this group.

6.9 Borates

The borates contain triangular BO_3^{3-} or tetrahedral BO_4^{5-}, and commonly both coordinations may occur in the same mineral. A common example is borax, $(Na_2B^{III}_2B^{IV}_2O_5(OH)_4\ 8H_2O)$.

6.10 Silicates

This group of minerals contains SiO_4^{4-} as the dominant polyanion. In these minerals the Si^{4+} cation is always surrounded by 4 oxygens in the form of a tetrahedron. Because Si and O are the most abundant elements in the Earth, this is the largest group of minerals and is divided into subgroups based on the degree of polymerization of the SiO_4 tetrahedra.

6.10.1 Orthosilicates

These minerals contain isolated SiO_4^{4-} polyanionic groups in which the oxygens of the polyanion are bound to one Si atom only, i.e., they are not polymerized. Examples are forsterite (Mg-olivine, Mg_2SiO_4), and pyrope (Mg-garnet, $Mg_3Al_2Si_3O_{12}$).

6.10.2 Sorosilicates

These minerals contain double silicate tetrahedra in which one of the oxygens is shared with an adjacent tetrahedron, so that the polyanion has formula $(Si_2O_7)^{6-}$. An example is epidote $(Ca_2Al_2FeO(OH)SiO_4\ Si_2O_7)$, a mineral common in metamorphic rocks.

6.10.3 Cyclosilicates

These minerals contain typically six-membered rings of silicate tetrahedra with formula $(Si_6O_{17})^{10-}$. An example is tourmaline.

6.10.4 Chain silicates

These minerals contain SiO_4 polyhedra that are polymerized in one direction to form chains. They may be single chains, so that of the four oxygen coordinating the Si atom, two are shared with adjacent tetrahedra to form an infinite chain with formula $(SiO_3)^{2-}$. The single chain silicates include the pyroxene and pyroxenoid minerals which are common constituents of igneous rocks. Or they may form double chains with formula $(Si_4O_{11})^{8-}$, as in the amphibole minerals, which are common in metamorphic rocks.

6.10.5 Sheet silicates

These minerals contain SiO_4 polyhedra that are polymerized in two dimensions to form sheets with formula $(Si_4O_{10})^{4-}$. Common examples are the micas in which the cleavage reflects the sheet structure of the mineral.

6.10.6 Framework silicates

These minerals contain SiO_4 polyhedra that are polymerized in three dimensions to form a framework with formula (SiO_2). Common examples are quartz (SiO_2) and the feldspars $(NaAlSi_3O_8)$ which are the most abundant minerals in the Earth's crust. In the feldspars Al^{3+} may substitute for Si^{4+} in the tetrahedra, and the resulting charge imbalance is compensated by an alkali cation (Na or K) in interstices in the framework.

7. The classification of crystals

The descriptive terminology of the discipline of crystallography is applied to crystals in order to describe their structure, symmetry, and shape. This terminology describes the crystal lattice, which provides a mineral with its ordered internal structure. It also describes and analyzes various types of symmetry. By considering what type of symmetry a mineral species possesses, the species may be categorized as a member of one of six crystal systems and one of thirty-two crystal classes.

The concept of symmetry describes the periodic repetition of structural features. Two general types of symmetry exist. These include translational symmetry and point symmetry. Translational symmetry describes the periodic repetition of a motif across a length or through an area or volume. Point symmetry, on the other hand, describes the periodic repetition of a motif about a single point. Reflection, rotation, inversion, and rotoinversion are all point symmetry operations.

A specified motif which is translated linearly and repeated many times will produce a lattice. A lattice is an array of points which define a repeated spatial entity called a unit cell. The unit cell of a lattice is the smallest unit which can be repeated in three dimensions in order to construct the lattice.

The number of possible lattices is limited. In the plane only five different lattices may be produced by translation. The French crystallographer Auguste Bravais (1811-1863) established that in three-dimensional space only fourteen different lattices may be constructed. These fourteen different lattices are thus termed the Bravais lattices.

The reflection, rotation, inversion, and rotoinversion symmetry operations may be combined in a variety of different ways. There are thirty-two possible unique combinations of symmetry operations. Minerals possessing the different combinations are therefore categorized as members of thirty-two crystal classes. In this classificatory scheme each crystal class corresponds to a unique set of symmetry operations. Each of the crystal classes is named according to the variant of a crystal form which it displays. Each crystal class is grouped as one of the six different crystal systems according to which characteristic symmetry operation it possesses.

A crystal form is a set of planar faces which are geometrically equivalent and whose spatial positions are related to one another by a specified set of symmetry operations. If one face of a crystal form is defined, the specified set of point symmetry operations will determine all of the other faces of the crystal form. A simple crystal may consist of only a single crystal form. A more complicated crystal may be a combination of several different forms. Example crystal forms are the parallelohedron, prism, pyramid, trapezohedron, rhombohedron and tetrahedron.

Each crystal class is a member of one of six crystal systems. These include the isometric, hexagonal, tetragonal, orthorhombic, monoclinic, and triclinic crystal systems. Every crystal of a certain crystal system shares a characteristic symmetry element - for example, a certain axis of rotational symmetry - with the other members of its system. The crystal system of a mineral species may sometimes be determined by examining a particularly well-formed crystal of the species (Nesse, 2004).

8. The economic value of minerals

Minerals that are of economic value can be classified as metallic or nonmetallic. Metallic minerals are those from which valuable metals (e.g. iron, copper) can be extracted for commercial use. Metals that are considered geochemically abundant occur at crustal abundances of 0.1 percent or more (e.g. iron, aluminum, manganese, magnesium, titanium). Metals that are considered geochemically scarce occur at crustal abundances of less than 0.1 percent (e.g. nickel, copper, zinc, platinum metals). Some important metallic minerals are: hematite (a source of iron), bauxite (a source of aluminum), sphalerite (a source of zinc) and galena (a source of lead). Metallic minerals occasionally but rarely occur as a single element (e.g. native gold or copper).

Nonmetallic minerals are valuable, not for the metals they contain, but for their properties as chemical compounds. Because they are commonly used in industry, they are also often referred to as industrial minerals. They are classified according to their use. Some industrial minerals are used as sources of important chemicals (e.g. halite for sodium chloride and borax for borates). Some are used for building materials (e.g. gypsum for plaster and kaolin for bricks). Others are used for making fertilizers (e.g. apatite for phosphate and sylvite for potassium). Still others are used as abrasives (e.g. diamond and corrundum).

8.1 Mineral deposits

Minerals are everywhere around us. For example, the ocean is estimated to contain more than 70 million tons of gold. Yet, it would be much too expensive to recover that gold because of its very low concentration in the water. Minerals must be concentrated into deposits to make their collection economically feasible. A mineral deposit containing one or more minerals that can be extracted profitably is called an ore. Many minerals are commonly found together (e.g. quartz and gold; molybdenum, tin and tungsten; copper, lead and zinc; platinum and palladium). Because various geologic processes can create local enrichments of minerals, mineral deposits can be classified according to the concentration process that formed them. The five basic types of mineral deposits are: hydrothermal, magmatic, sedimentary, placer and residual.

Hydrothermal mineral deposits are formed when minerals are deposited by hot, aqueous solutions flowing through fractures and pore spaces of crustal rock. Many famous ore bodies have resulted from hydrothermal depositon, including the tin mines in Cornwall, England and the copper mines in Arizona and Utah, USA. Magmatic mineral deposits are formed when processes such as partial melting and fractional crystallization occur during the melting and cooling of rocks.

Pegmatite rocks formed by fractional crystallization can contain high concentrations of lithium, beryllium and cesium. Layers of chromite (chrome ore) were also formed by igneous processes in the famous Bushveld Igneous Complex in South Africa.

Several mineral concentration processes involve sedimentation or weathering. Water soluble salts can form sedimentary mineral deposits when they precipitate during evaporation of lake or seawater (evaporite deposits). Important deposits of industrial minerals were formed in this manner, including the borax deposits at Death Valley and Searles Lake, and the marine deposits of gypsum found in many states.

Minerals with a high specific gravity (e.g. gold, platinum, diamonds) can be concentrated by flowing water in placer deposits found in stream beds and along shorelines. The most famous gold placer deposits occur in the Witwatersrand basin of South Africa. Residual mineral deposits can form when weathering processes remove water soluble minerals from an area, leaving a concentration of less soluble minerals. The aluminum ore, bauxite, was originally formed in this manner under tropical weathering conditions. The best known bauxite deposit in the United States occurs in Arkansas.

8.2 Mineral utilization

Minerals are not evenly distributed in the earth's crust. Mineral ores are found in just a relatively few areas, because it takes a special set of circumstances to create them. Therefore, the signs of a mineral deposit are often small and difficult to recognize. Locating deposits requires experience and knowledge. Geologists can search for years before finding an economic mineral deposit. Deposit size, its mineral content, extracting efficiency, processing costs and market value of the processed minerals are all factors that determine if a mineral deposit can be profitably developed. For example, when the market price of copper increased significantly in the 1970s, some marginal or low-grade copper deposits suddenly became profitable ore bodies. After a potentially profitable mineral deposit is located, it is mined by one of several techniques. Which technique is used depends upon the type of deposit and whether the deposit is shallow and thus suitable for surface mining or deep and thus requiring sub-surface mining.

Surface mining techniques include: open-pit mining, area strip mining, contour strip mining and hydraulic mining. Open-pit mining involves digging a large, terraced hole in the ground in order to remove a near-surface ore body. This technique is used in copper ore mines in Arizona and Utah and iron ore mines in Minnesota, USA. Area strip mining is used in relatively flat areas. The overburden of soil and rock is removed from a large trench in order to expose the ore body. After the minerals are removed, the old trench is filled and a new trench is dug. This process is repeated until the available ore is exhausted. Contour strip mining is a similar technique except that it is used on hilly or mountainous terrains. A series of terraces are cut into the side of a slope, with the overburden from each new terrace being dumped into the old one below.

Hydraulic mining is used in places such as the Amazon in order to extract gold from hillsides. Powerful, high-pressure streams of water are used to blast away soil and rock containing gold, which is then separated from the runoff. This process is very damaging to the environment, as entire hills are eroded away and streams become clogged with sediment. If land subjected to any of these surface mining techniques is not properly

restored after its use, then it leaves an unsightly scar on the land and is highly susceptible to erosion.

Some mineral deposits are too deep to be surface mined and therefore require a sub-surface mining method. In the traditional sub surface method a deep vertical shaft is dug and tunnels are dug horizontally outward from the shaft into the ore body. The ore is removed and transported to the surface. The deepest such subsurface mines (deeper than 3500 m) in the world are located in the Witwatersrand basin of South Africa, where gold is mined. This type of mining is less disturbing to the land surface than surface mining. It also usually produces fewer waste materials. However, it is more expensive and more dangerous than surface mining methods.

A newer form of subsurface mining known as in-situ mining is designed to coexist with other land uses, such as agriculture. An in-situ mine typically consists of a series of injection wells and recovery wells built with acid-resistant concrete and polyvinyl chloride casing. A weak acid solution is pumped into the ore body in order to dissolve the minerals. Then, the metal-rich solution is drawn up through the recovery wells for processing at a refining facility. This method is used for the in-situ mining of copper ore.

Once an ore has been mined, it must be processed to extract pure metal. Processes for extracting metal include smelting, electrowinning and heap leaching. In preparation for the smelting process, the ore is crushed and concentrated by a flotation method. The concentrated ore is melted in a smelting furnace where impurities are either burned-off as gas or separated as molten slag. This step is usually repeated several times to increase the purity of the metal. For the electrowinning method ore or mine tailings are first leached with a weak acid solution to remove the desired metal. An electric current is passed through the solution and pure metal is electroplated onto a starter cathode made of the same metal. Copper can be refined from oxide ore by this method. In addition, copper metal initially produced by the smelting method can be purified further by using a similar electrolytic procedure. Gold is sometimes extracted from ore by the heap leaching process. A large pile of crushed ore is sprayed with a cyanide solution. As the solution percolates through the ore it dissolves the gold. The solution is then collected and the gold extracted from it. All of the refining methods can damage the environment. Smelters produce large amounts of air pollution in the form of sulfur dioxide which leads to acid rain. Leaching methods can pollute streams with toxic chemicals that kill wildlife (Roberts, Campbell & Rapp, 1990).

8.3 Mineral sufficiency and the future

Mineral resources are essential to life as we know it. A nation cannot be prosperous without a reliable source of minerals, and no country has all the mineral resources it requires. The United States has about 5 percent of the world's population and 7 percent of the world's land area, but uses about 30 percent of the world's mineral resources. It imports a large percentage of its minerals; in some cases sufficient quantities are unavailable in the U.S., and in others they are cheaper to buy from other countries. Certain minerals, particularly those that are primarily imported and considered of vital importance, are stockpiled by the United States in order to protect against embargoes or other political crises. These strategic minerals include: bauxite, chromium, cobalt, manganese and platinum.

Because minerals are produced slowly over geologic time scales, they are considered non-renewable resources. The estimated mineral deposits that are economically feasible to mine are known as mineral reserves. The growing use of mineral resources throughout the world raises the question of how long these reserves will last. Most minerals are in sufficient supply to last for many years, but a few (e.g. gold, silver, lead, tungsten and zinc) are expected to fall short of demand in the near future. Currently, reserves for a particular mineral usually increase as the price for that mineral increases. This is because the higher price makes it economically feasible to mine some previously unprofitable deposits, which then shifts these deposits to the reserves. However, in the long term this will not be the case because mineral deposits are ultimately finite.

There are ways to help prolong the life of known mineral reserves. Conservation is an obvious method for stretching reserves. If you use less, you need less. Recycling helps increase the amount of time a mineral or metal remains in use, which decreases the demand for new production. It also saves considerable energy, because manufacturing products from recycled metals (e.g. aluminum, copper) uses less energy than manufacturing them from raw materials. As a result, mineral prices are kept artificially low which discourages conservation and recycling.

9. References

Blackburn, W.H., Dennen, W.H. (1988). *Principles of Mineralogy*. (1st edition), Wm.C. Brown Publishers, ISBN 069715078X, Dubuque, Iowa.

Deer, W.A., Howie, R.A., Zussman, J. (1992). *An Introduction to the Rock Forming Minerals*. (2nd edition), ISBN 0-582-30094-0, Longman Publishing Co, London.

Klein, C., Hurlbut, Jr.C.S. (1985). *Manual of Mineralogy*. (20th edition), John Wiley & Sons, ISBN 047180580, New York.

Lutgens, F.K. and Tarbuck, E.J. (2000). *Essentials of Geology*. (7th edition), Prentice Hall, ISBN, 0130145440, New York.

Nesse, W.D. (2000). *Introduction to mineralogy*. Oxford University Press, ISBN-10: 0195106911; New York.

Nesse, W. D. (2004). *Introduction to Optical Mineralogy*. Oxford University Press, ISBN 019522132X, New York.

Roberts, W.L., Campbell, T.J., Rapp, Jr. G.R. (1990). *Encyclopedia of Minerals*. (2nd edition), Van Nostrand, Reinhold, New York.

Rudnick, R.L., Fountain, M.D. (1995). Nature and composition of the continental crust: A lower crustal perspective. *Reviews of geophysics*, Vol. 33, No. 3, pp. 267-309.

Zoltai, T., Stout, J.H. (1984). *Mineralogy: Concepts and Principles*. Burgess Publishing Co, ISBN 9780024320100, Minneapolis.

Cation Distribution and Equilibration Temperature of Amphiboles from the Sittampundi Complex, South India

B. Maibam[1] and S. Mitra [2]
[1]Department of Earth Sciences, Manipur University, Canchipur,
[2]Department of Geological Sciences, Jadavpur University,
India

1. Introduction

Amphiboles are utilized as indicators of temperature, pressure, volatile content, and oxidation state by the petrologist; while to a mineralogist the amphiboles provide instructive examples of structural phase transitions, cation ordering over multiple crystallographic sites, and a variety of atomic substitutions (Ghiorso & Evans, 2002). Relating the chemistry of calcic-amphibole to the *P-T* conditions of crystallization is a daunting task, because the amphibole crystal structure consists of a variety of cation sites, accommodating elements of a wide range of ionic radii and valences.

Amphibole has the general formula $A_{0-1}X_2Y_5Z_8O_{22}(OH,F,Cl)_2$, where A = Na, K; X = Na, Li, Ca, Mn, Fe^{2+}, Mg; Y = Mg, Fe^{2+}, Mn, Al, Fe^{3+}, Ti, Li; Z = Si, Al. Calcic amphiboles crystallize in the monoclinic crystal system (space group $C2/m$) (Hawthorne, 1983). An essential feature of amphibole is the presence of $(Si,Al)O_4$ tetrahedra linked to form chain which has double the width of those in pyroxenes. The four octahedral sites (M1, M2, M3 and M4) present in amphibole are neither symmetric nor of equal energy. The M1, M2 and M3 octahedra, which accommodate Mn, Fe^{2+}, Mg, Fe^{3+}, Cr, Al and Ti cations, share edges to form octahedral bonds parallel to *c*-axis. The M1 and M3 (regular octahedra) are co-ordinated by four oxygens and two (OH,F) groups. The M2 (slightly distorted octahedra) is co-ordinated by six oxygens. The cations Ti, Al and Fe^{3+} usually prefer the M2 site, especially when Na occupies the adjacent M4 site (e.g., Papike et al., 1969; Papike, 1988). The M4 site, which accommodates Na, Li, Ca, Mn, Fe^{2+}, Mg cations, is commonly a slightly-distorted octahedron when occupied by Mn, Fe^{2+} and Mg; and it is eight co-ordinated when occupied by Na or Ca (Papike, 1988). Two independent chains of tetrahedral sites, T1 and T2 cross-link the octahedral strips. The T1 site shares three oxygens with other tetrahedra and is more regular and smaller than T2. The T2 site shares two oxygens with adjacent tetrahedra. Papike et al. (1969) observed the preference of Al for the T1 site relative to the T2 site in hornblende, based on <T1–O> and <T2–O> lengths. The large and highly distorted A – polyhedron occurs between the tetrahedral chain. This site can be vacant, partially filled, or fully occupied by Na and/or K in $C2/m$ structure.

The present work was undertaken to investigate the chemical characteristic of calcic and magnesian amphiboles from upper amphibolite to granulite facies rocks in the Sittampundi complex. In addition, main objective of this work is to characterize iron ions of different valence states and their distribution within the crystal lattice, both in calcic and magnesian amphiboles of amphibolite, meta-anorthosite and chromitite respectively, using ^{57}Fe Mössbauer spectroscopy and estimate the equilibration temperature attained by the host rock.

2. Geological setting

Sittampundi Anorthosite Complex occurs as a layered igneous body. The area forms a part of the granulite terrain of South India. Major rock types are chromitite bearing meta-anorthosite, amphibolite, basic granulite, two pyroxene granulite, leptynite, biotite gneiss and pink granite. In this complex, amphibolite and chromitite occur as small discontinuous bands/lenses within the meta-anorthosite. Basic granulite occurs as discontinuous but conformable layers within meta-anorthosite. Two-pyroxene granulite and pink granite occur within the biotite-gneiss. The former two are the later intrusive phases (Subramanium, 1956).

In the study area amphiboles commonly occur throughout the complex as mosaic aggregates, rarely as inclusion in plagioclase. The mineral assemblages of the major amphibole-bearing rocks are as follows:

i.   Amphibolite: calcic amphibole + plagioclase ± garnet ± cpx ± chlorite ± sphene ± opaque
ii. Meta-anorthosite: plagioclase ± calcic amphibole ± anthophyllite ± clinozoisite ± clinopyroxene ± garnet ± corundum ± epidote ± scapolite ± carbonate ± hercynite ± spinel ± sillimanite ± muscovite ± chlorite ± sphene ± opaque.
iii. Chromitite: chromite ± calcic amphibole ± anthophyllite ± rutile ± chlorite.

In the basic granulite and two-pyroxene granulite, calcic amphibole occurs as rim surrounding diopside. In meta-anorthosite, tschermakitic hornblende forms triple junction with plagioclase; and triple junctions are common between amphibole crystals in amphibolite.

3. Experimental methods

Electron microprobe analysis of the samples has been obtained by JEOL-733 superprobe (wavelength dispersive method) at 15 kV with beam current of 0.01 µA and 2 µm beam diameter and Cameca CAMEBAX SX 50 probe with digital computer PDP 11.53. Operating condition was 15 kV accelerating voltage with a beam current of 15 nA and beam size ~1µm. The average spectrum count was compared with the natural standards.

The room temperature (RT) Mössbauer spectra were recorded in a conventional constant acceleration spectrometer (Wissel GmbH, Germany) with a 25 mCi Co/Rh source. Two mirror-imaged spectra were obtained in 512 out of 1024 channels. The finely powdered samples were pressed tight in plexi holders, and based on the iron content (known from EPMA) determined amounts of the samples were taken to optimise the thickness effect. The amount used for the samples are: 60 mg (sample 3c), 68 mg (sample 15b) and 92 mg (sample 30a). The velocity calibration was performed with respect to pure metallic iron (99.99%). The spectra were then fitted to the sum of Lorentzian functions with a non-linear least square fitting programme of Meerwall (1975) on an IBM personal computer. For each doublet the intensity and line-width of the peaks were constrained to be equal.

4. Results and discussion

4.1 Chemical composition of Sittampundi amphiboles

Amphiboles have been classified according to Leake (1978). For this classification empirical Fe^{3+} in calcic amphibole was estimated from EPMA chemistry following Schumacher (1991). Site occupancy of different elements in amphiboles (viz. calcic amphiboles, anthophyllite) was assigned on the basis of 23 oxygens following Papike (1988). We present here the composition of only three amphiboles (3c, 15b, 30a), the chemistry of which were determined employing both the EPMA and Mössbauer spectroscopy (Table 1).

Oxides	3c	15b	30a
SiO_2	42.49	42.48	56.44
TiO_2	1.26	0.28	0.07
Cr_2O_3	0.06	0.03	0.48
Al_2O_3	11.47	16.82	2.86
FeO	18.78	10.80	7.80
MnO	0.34	0.15	0.20
MgO	9.04	13.09	28.35
NiO	0.01	-	0.11
CaO	11.57	11.52	0.60
Na_2O	1.91	2.52	0.30
K_2O	0.54	0.08	-
Total	97.47	97.57	97.21
Number of cations on the basis of 23 oxygen			
Si	6.459	6.145	7.699
Fe^{3+}	0.402	0.263	0.065
Al^{IV}	1.139	1.592	0.236
Sum (T)	8.000	8.000	8.000
Al^{VI}	0.916	1.290	0.224
Ti	0.144	0.030	0.007
Cr	0.007	0.003	0.052
Fe^{3+}	0.317	0.182	0.166
Fe^{2+}	1.293	0.675	0.368
Mg	2.048	2.820	4.148
Mn	0.044	-	0.023
Ni	0.001	-	0.012
Sum (M123)	4.770	5.000	5.000
Fe^{2+}	0.377	0.193	0.290
Mg	-	0.015	1.618
Number of cations on the basis of 23 oxygen			
Mn	-	0.018	-
Ca	1.884	1.794	0.087
Na	-	-	0.005
Sum (M4)	2.261	2.020	2.000
Na	0.502	0.709	0.075
K	0.104	0.014	-
Sum (A)	0.660	0.723	0.075

Table 1. Chemical composition and cation distribution of amphiboles studied by EPMA and Mössbauer spectroscopy

4.2 Characterization of samples

Studied amphiboles were selected from different metamorphic rocks, such as amphibolite (sample 3c), meta-anorthosite (sample 15b), chromitite (sample 17c, 18b, 25, 29, 30a) and basic granulite (sample 6b, 13). Mössbauer analysis of two calcic amphiboles (sample 3c and 15b) and one anthophyllite (sample 30a) were considered in the present study. Sample 3c was taken from amphibolite occurring in the eastern part of the Sittampundi complex, near the Cauvery river bed of Kettuvelanpalaiyam. Sample 15b was taken from meta-anorthosite in the western side of Pamandapalaiyam, near the road between Cholasiramani and Tottiyamthottam. Sample 30a was taken from chromitite in the western side of Karungalpatti. The crystal chemistry of the last three amphiboles was determined from microprobe analyses combined with the results of Mössbauer spectroscopy. The studied amphibole species present in different rock types are identified as follows:

i.   Meta-anorthosite: ferroan pargasite
ii.   Chromitite: edenitic hornblende, magnesio-hornblende, tchermakitic hornblende, anthophyllite
iii. Basic granulite: edenite, edenitic hornblende, pargasite, pargasitic hornblende, ferro-hornblende, actinolite
iv.   Two-pyroxene granulite: edenitic hornblende, magnesio-hornblende
v.   Amphibolite: ferroan pargasitic hornblende

The variational range of X_{Mg} [Mg/(Mg+Fe)], X_{Ca} [Ca/(Ca+Mg)], X_K [K/(K+Na)], TiO_2 and Cr_2O_3 of calcic amphiboles and anthophyllite in different rock types are given in Table 2. TiO_2 content of calcic amphibole in meta-anorthosite is low (0.21-0.36 wt%) but it is high in some basic granulite ($\cong 2.07$ wt%). TiO_2 content of calc amphiboles in two-pyroxene granulite, amphibolite and chromitite range between 0.90-1.42 wt%, 1.02-1.47 wt%, 0.47-1.09 wt%, respectively. Cr_2O_3 content of calcic-amphiboles and anthophyllite in chromitite show wide variation (0.00-1.12 and 0.38-0.83 wt%, respectively), but in other rock types it is low (<0.34 wt%). The plot of Si vs. (Ca+Na+K) of Sittampundi calcic amphiboles, following Shido & Miyashiro (1959) and Binns (1965), show a cluster in the field of green and brown hornblende (belonging to the high amphibolite and granulite facies). In anthophyllite, the ratio of (Na+K) to tetrahedral Al is remarkably constant and it lies close to ¼. Thus, for each 0.1 Na ion per formula unit, there are 0.4 tetrahedral Al ions substituting for Si.

Amphiboles	Host rock	X_{Mg}	X_{Ca}	X_K	TiO_2	Cr_2O_3
Calcic-amphibole	Basic granulite	0.36-0.89	0.29-0.55	0.00-0.30	0.13-2.07	0.02-0.33
	Two pyroxene granulite	0.68-0.78	0.33-0.37	0.00-0.02	0.90-1.42	0.00-0.33
	Anorthosite	0.68-0.69	0.38-0.40	0.02-0.11	0.21-0.36	0.00-0.35
	Amphibolite	0.45-0.47	0.47-0.49	0.14-0.17	1.02-1.47	0.02-1.12
Anthophyllite	Chromitite	0.84-0.87	0.01-0.02	0.02-0.05	0.03-0.14	0.38-0.83

Table 2. The range of the values of X_{Mg}, X_{Ca}, X_K, TiO_2 and Cr_2O_3 for calcic amphiboles and anthophyllite in the studied samples based on the EPMA

4.3 Mössbauer study and assignment

The room temperature (298 K) Mössbauer spectrum of the calcic amphiboles and anthophyllite show a close range of isomer shift (IS) of octahedral Fe^{2+} ion (Table 3) and the assignment is made from the quadrupole splitting (QS) values. The quadrupole splitting of M1, M2 and M3 sites which are regular octahedra, completely overlap in the case of Mg-Fe amphibole, but chemistry becoming more complex with Ca, Na, Al and Fe^{3+} the QS values tend to decrease (Goldman, 1979).

Sample No	IS (mm/s)	QS (mm/s)	Distribution	Area (%)	Width (mm/s)	χ^2	$Fe^{3+}/\Sigma Fe$
Calcic-amphibole							
	1.23	2.93	$Fe^{2+}(M1M3)$	33.59	0.35		
	1.25	2.47	$Fe^{2+}(M2)$	20.52	0.41		
3c	1.47	1.41	$Fe^{2+}(M4)$	15.77	0.49	1.38	0.30
	0.41	0.82	$Fe^{3+}(M2)$	13.29	0.37		
	0.02	0.35	$Fe^{3+}(T)$	16.83	0.38		
	1.22	2.84	$Fe^{2+}(M1M3)$	32.05	0.35		
	1.25	2.35	$Fe^{2+}(M2)$	19.35	0.44		
15b	1.42	1.40	$Fe^{2+}(M4)$	14.75	0.41	3.00	0.20
	0.47	0.86	$Fe^{3+}(M2)$	13.83	0.41		
	0.07	0.38	$Fe^{3+}(T)$	20.02	0.41		
Anthophyllite							
	1.28	2.88	$Fe^{2+}(M1M3)$	33.51	0.38		
	1.13	2.42	$Fe^{2+}(M2)$	7.89	0.39		
30a	1.26	1.93	$Fe^{2+}(M4)$	32.60	0.40	4.94	0.26
	0.50	0.75	$Fe^{3+}(M2)$	18.71	0.45		
	0.01	0.55	$Fe^{3+}(T)$	7.29	0.25		

Table 3. Mössbauer parameters of amphiboles with $Fe^{3+}/\Sigma Fe$ ratio

The M1 and M3 sites of the calcic amphibole are usually occupied by the Fe^{2+} and Mg whereas the M2 site is occupied by Fe^{3+}, Al^{3+}, Fe^{2+} and Mg (Skogby, 1987). With incorporation of more Al and Fe^{3+} in amphibole structure the M–O bond length of M2 site become smaller (Papike et al., 1969) and the QS for Fe^{2+} in the M2 site would manifest a reduction compared to that of the M1 and M3 sites. However, Della Ventura et al. (2005) showed that because the M2 site is adjacent to the M4 site, the QS value change correlated directly with the Li–Na substitution. In aluminous amphibole Fe^{2+} is enriched in M3 site relative to M1, which host Mg. In calcic amphibole, the M4 octahedral site is most distorted and is preferentially occupied by larger cation such as Ca, Na, Mn although occurrence of Fe^{2+} and Mg at M4 site has been postulated by some authors (Goldman & Rossman, 1977; Goldman, 1979; Skogby & Annersten, 1985; Skogby, 1987). Difficulty in determining Fe^{2+} at M4 site arises because Ca occupies 85-95% of this site (Leake, 1968). However, a high degree of distortion of the M4 site is expected to produce large Fe^{2+} crystal field splitting analogous to that seen in the distorted M2 site of orthopyroxene (Goldman & Rossman, 1977).

4.3.1 Calcic amphiboles

The room temperature (298 K) Mössbauer spectra (Figs. 1a, b) of the ferroan pargasitic hornblende (from amphibolite, sample 3c) and ferroan pargasite (from meta-anorthosite, sample 15b), show two asymmetric paramagnetic doublets of which the more intense peak lies at lower velocity region ($\cong -0.13$ mm/s). The fitting of the Mössbauer spectrum was approached in two ways: viz. (a) four doublet fit (three doublets for octahedral Fe^{2+} ions, and the other for octahedral Fe^{3+}), and (b) five doublet fit (three doublets for octahedral Fe^{2+} ions, and two for octahedral and tetrahedral Fe^{3+}). The five doublet fit improves the chi-square value and narrows down the line-width (0.35-0.49 mm/s) from wider line widths of the four doublet fit. The Mössbauer parameters as well as the assignment of the doublets are presented in Table 3. The basis of assignment is stated below.

In five doublet fit, the outermost doublets (1-1) with IS 1.23 mm/s (for sample 3c) and 1.22 mm/s (for sample 15b) and corresponding QS values of 2.93 mm/s, 2.84 mm/s are assigned to Fe^{2+} in M1 and M3 sites (Goldman, 1979; Schumacher, 1991). However, the IS values (at RT) for silicate phases range from 1.3 to 1.43 mm/s for octahedral Fe^{2+} (vide Mitra, 1992, Table 1.3). The doublets (3-3) showing IS 1.25 mm/s (for sample 3c), 1.25 mm/s (for sample 15b) and the corresponding QS 2.47 mm/s, 2.35 mm/s are assigned to Fe^{2+} in the M2 site (Goldman, 1979; Schumacher, 1991). The innermost doublets (5-5) with IS 1.47 mm/s (for sample 3c), 1.42 m/s (for sample 15b) and corresponding QS 1.41 mm/s, 1.40 mm/s are assigned to Fe^{2+} in the M4 site (Goldman, 1979; Skogby & Annersten, 1985; Skogby, 1987; Schumacher, 1991). Differing QS value arises from difference in magnetic field generated by the orbital momentum of Fe^{2+} ion in those octahedral (M) sites giving rise to different orbital contributions H_{orb} and H_{hf} and also to the differences in covalency (Linares et al., 1983). The M4 site is bonded more covalently with Ca ion, with consequent decrease in the QS value of Fe^{2+} at this site. The decrease can also be accounted by considering the NNN effect where one Fe^{2+} is substituted by cations such as Mg^{2+} or Ca^{2+} and the super exchange in Fe-O-Fe bond, between octahedral sites (Goodenough, 1963). The entry of Al^{3+} ion in the M2 site could also affect the QS value of the M4 site. The lower QS value of 1.41 mm/s for sample 3c (Al^{VI} = 0.916 p.f.u.) and 1.40 mm/s for sample 15b (Al^{VI} = 1.290 p.f.u.) of M4 site may also signify greater distortion caused by faster rate of its formation (Jenkins, 1987; Skogby, 1987) during cooling.

4.3.1.1 Fe^{3+} in calcic amphiboles

The doublets (4-4) showing IS 0.41 mm/s (for sample 3c), 0.47 mm/s (for sample 15b) and corresponding QS 0.82 mm/s, 0.86 mm/s are assigned to Fe^{3+} in the M2 site (Goldman, 1979; Skogby & Annersten, 1985; Skogby, 1987; Schumacher, 1991). Fe^{3+} at distorted polyhedron shows larger QS than the regular one. So the larger QS value of Fe^{3+} doublet represents its occupancy at distorted M2 site.

The IS value of the other Fe^{3+} doublet (2-2) suggests a tetrahedral co-ordination rather than Fe^{3+} at octahedral site. So the doublets with IS 0.02 mm/s (for sample 3c), 0.07 mm/s (for sample 15b) and the corresponding QS 0.35 mm/s, 0.38 mm/s are assigned to Fe^{3+} in the less-distorted tetrahedral site. Extremely low value of IS for sample 3c is due to Fe^{3+} at an almost regular tetrahedron.

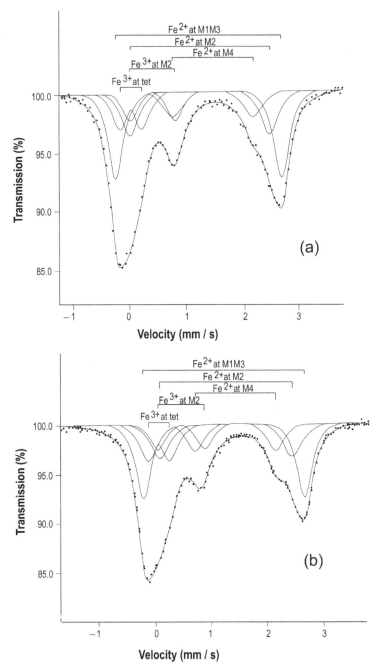

Fig. 1. Room temperature ^{57}Fe Mössbauer spectra of calcic amphiboles (a) sample 3c (b) sample 15b (Dots: Observed MS counts; continuous lines: fit with Lorentzian line shape)

The structural formulae of calcic amphiboles determined from EPMA and Mössbauer analysis are

Sample 3c

A	M4	M1M2M3	T

$(Na_{0.502}K_{0.104})(Fe^{2+}_{0.377}Ca_{1.884})(Al_{0.916}Ti_{0.144}Cr_{0.007}Fe^{3+}_{0.317}Fe^{2+}_{1.293}Mg_{2.048}Mn_{0.044}Ni_{0.001})(Fe^{3+}_{0.402}Al_{1.139}Si_{6.459})O_{23}$

Sample 15b

$(Na_{0.709}K_{0.014})(Fe^{2+}_{0.193}Ca_{1.794}Mg_{0.015}Mn_{0.018})(Al_{1.290}Ti_{0.030}Cr_{0.003}Fe^{3+}_{0.182}Fe^{2+}_{0.675}Mg_{2.820})(Fe^{3+}_{0.263}Al_{1.592}Si_{6.145})O_{23}$

The Fe^{2+} content (15.76% of total iron) of M4 site for sample 3c (total iron content is greater than sample 15b) determined by spectral fitting gives oversaturation of M4 site with vacancies in M1, M2 and M3 sites. The reduction of Fe^{2+} at M4 site in computer fitting causes increase in chi-square value of fitting which is not agreeable. So the enrichment of Fe^{2+} at highly distorted M4 site (QS = 1.41 mm/s) is evidently related to the substitutional Fe^{3+} at tetrahedral site. This coupled entry of Fe^{2+} at M4 site and Fe^{3+} at tetrahedral site point to a higher ordering by slow cooling. However, in natural calcic amphibole cation, M4 site totaling greater than 2.00 p.f.u. is not uncommon (Deer et al., 1978). This type of over saturation (in M4 site) and vacancies (in M1, M2 and M3 sites) are present in sample 3c (Table 1).

4.3.2 Anthophyllite

The room temperature (298 K) Mössbauer spectrum (Fig. 1c) of anthophyllite occurring in chromitite (sample 30a) consists of two asymmetric, paramagnetic doublets with more intense peak at lower velocity region ($\cong 0.25$ mm/s). The inner, more intense doublet can, on the basis of the crystal structure determination, be assigned to Fe^{2+} in M4; and the outer doublet to Fe^{2+} in M1, M2 and M3 sites (Seifert & Virgo, 1973). Similar to the calcic amphiboles, the fitting of Mössbauer spectrum was approached likewise in two ways: (a) four doublet fit and (b) five doublet fit. The five doublet fit here again showed better Mössbauer parameter (χ^2) than those of the four doublet fit. The Mössbauer parameter as well as the assignment of the iron doublet to specific crystallographic site of anthophyllite, following calcic amphibole assignment, is presented in Table 3. The distribution of Fe^{2+}, Fe^{3+} in the different sites of anthophyllite (sample 30a) determined by spectral fitting indicates the absence of vacancies in different sites. The structural formula of anthophyllite determined from EPMA and Mössbauer analysis is

A	M4	M1M2M3	T

$Na_{0.076}(Fe^{2+}_{0.290}Ca_{0.087}Mg_{1.618}Na_{0.005})(Al_{0.224}Ti_{0.007}Cr_{0.052}Fe^{3+}_{0.116}Fe^{2+}_{0.368}M_{4.148}Mn_{0.023}Ni_{0.012})(Fe^{3+}_{0.065}Al_{0.236}Si_{7.699})O_{23}$

The higher QS value (0.55 mm/s) of Fe^{3+} doublet (2-2) of tetrahedral site suggests that this site is more regular (as it should be) than the tetrahedral site of calcic amphiboles (0.35 mm/s for sample 3c and 0.38 mm/s for sample 15b).

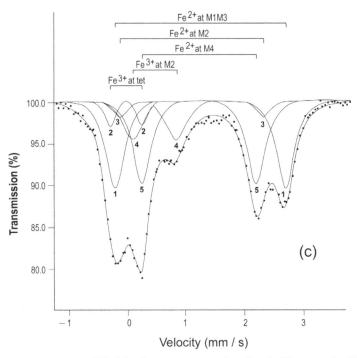

Fig. 1. (c) Room temperature ^{57}Fe Mössbauer spectra of anthophyllite sample (30a, Dots: Observed MS counts; continuous lines: fit with Lorentzian line shape)

It should be mentioned that the chi-square values of samples 15b (χ^2 = 3.84) and 30a (χ^2 = 4.94) are not within the usual limits. X-ray study shows that the sample 15b is inhomegenous in character. However, the inhomogeneity could be seen in backscattered electron (BSE) images but the fineness of the lamellae evaded analysis. The spectral fitting of these samples show higher chi-square values, but the fitting logic and Mössbauer parameters are in conformity with the earlier reported values (e.g., Schumacher, 1991).

5. Equilibration temperature of the amphiboles

Mineral assemblage of the studied calcic amphibole bearing samples indicates that the rock underwent amphibolite to lower granulite facies metamorphism. Amphiboles are good indicators of temperature and pressure over wide range of metamorphic conditions (Shido & Miyashiro, 1959; Engel & Engel, 1962; Binns, 1965; Leake, 1965, 1971; Bard, 1970; Ernst, 2002). We have used Holland & Blundy's (1994) amphibole-plagioclase thermometry to calculate the equilibration temperature of the studied Sittampundi calcic amphiboles. In the studied samples, the magmatic Ca-amphibole probably equilibrated with calcic plagioclase but may not have been silica-saturated, and the mineralogical assemblages of the samples do not show any free quartz, so the thermometer (B) was utilized. A pressure of 10 kbar has been estimated using Newton & Perkin's (1982) geobarometry for the basic granulites and this is considered for calculating the equilibrium temperature of the studied samples. The

estimated equilibration temperatures of the calcic amphiboles are presented in Table 4. Due to the absence of suitable coexisting mineral pairs, for the anthophyllite associated with the Sittampundi chromitites, we used the Plyusnina's (1982) curve to calculate the equilibration temperature and found to range from 530-560ºC. We have also used Graham & Powell's (1984) hornblende-garnet geothermometry to calculate the equilibration temperature of the basic granulites. The temperatures calculated from the two methods are falling within the same range.

Rock type	Sample no.	Temperature (ºC)	Method
Basic granulite	11b	800	Holland & Blundy (1994)
		780-800	Graham & Powell (1984)
	12c	650-680	Graham & Powell (1984)
	33	695-750	Holland & Blundy (1994)
	41a	680-700	Graham & Powell (1984)
Two-pyroxene granulite	6b	540	Plyusnina (1982)
	13	540	Plyusnina (1982)
Meta-anorthosite	15b	890	Holland & Blundy (1994)
Amphibolite	3c	800-820	Holland & Blundy (1994)
Chromitite	17c	530	Plyusnina (1982)
	18b	560	Plyusnina (1982)
	25	550	Plyusnina (1982)
	29	560	Plyusnina (1982)
	30a	540	Plyusnina (1982)

Table 4. Equilibration temperatures of Sittampundi amphiboles

6. Acknowledgements

Sincere thanks are due to the Director of Bayerisches Geoinstitut, Bayreuth for allowing to use the Mössbauer spectrometer and Electron Probe Microanalyser at his institute during SM's visit. The authors are indebted to Prof. H.S. Moon of Yonsei University, Seoul, for providing EPMA analysis of other samples. We thank Prof. E.J. Essene for his insightful comments on an earlier version of the paper. Asoke Kumar Samanta is acknowledged for his help in the initial draft of this paper. Financial support by the University Grants Commission, New Delhi to SM is acknowledged.

7. References

Bard, J.B. (1970). Composition of hornblendes formed during the Hercynian progressive metamorphism of the Archaean metamorphic belt (SW Spain), *Contribution to Mineralogy and Petrology*, Vol. 28, pp. 117-134

Binns, R.A. (1965). The mineralogy of metamorphosed basic rocks from the Willyama complex, Broken Hill District, New South Wales, Part I: Hornblendes, *Mineralogical Magazine*, Vol. 35, pp. 306-326

Deer, W.A.; Howie, R.A. & Zussman, J. (1978). *Rock-forming Minerals: Single-chain Silicates*, Longman, London

Della Ventura, G.; Redhammer, G.J.; Iezzi, G.; Hawthorne, F.C.; Papin, A. & Robert, J.-L. (2005). A Mössbauer and FTIR study of synthetic amphiboles along the magnesioriebeckite – ferri-clinoholmquistite join, *Physics and Chemistry of Minerals*, Vol. 32, pp. 103-113

Engel, A.E.J. & Engel, C.G. (1962). Hornblendes formed during progressive metamorphism of amphibolites, northwest Adirondack Mountains, New York, *Bulletin Geological Society of America*, Vol. 73, pp. 1499-1514

Ernst, W.G. (2002). Paragenesis and thermobarometry of Ca-amphiboles in the Barcroft granodioritic pluton, central White Mountains, eastern California, *American Mineralogist*, Vol. 87, pp. 478-490

Ghiorso, M.S. & Evans, B.W. (2002). Thermodynamics of amphiboles: $Ca-Mg-Fe^{2+}$ quadrilateral, *American Mineralogist*, Vol. 87, pp. 79-98

Goldman, D.S. (1979). A re-evaluation of the Mössbauer spectroscopy of calcic amphiboles, *American Mineralogist*, Vol. 64, pp. 109-118

Goldman, D.S. & Rossman, G.R. (1977). The identification of Fe^{2+} in the M(4) site of calcic amphiboles, *American Mineralogist*, Vol. 62, pp. 205-216

Goodenough, J.B. (1963). *Magnetism and the Chemical Bond*, Interscience, New York

Graham, C.M. & Powell, R. (1984). A garnet-hornblende geothermometer: calibration, testing and application to the Pelona schist, Southern California, *Journal of Metamorphic Geology*, Vol. 2, pp. 13-31

Hawthorne, F.C. (1983). The crystal chemistry of the amphiboles, *Canadian Mineralogist*, Vol. 21, pp. 174-481

Holland, T.J.B. & Blundy, J. (1994). Non-ideal interactions in calcic amphiboles and their bearing on amphibole-plagioclase thermometry, *Contribution to Mineralogy and Petrology*, Vol. 116, pp. 433-447

Jenkins, D.M. (1987). Synthesis and characterization of tremolite in the system $H_2O-CaO-MgO-SiO_2$, *American Mineralogist*, Vol. 72, pp. 707-715

Leake, B.E. (1965). The relationship between tetrahedral aluminium and the maximum possible octahedral aluminium in natural calciferous and sub-calciferous amphiboles, *American Mineralogist*, Vol. 50, 843-851

Leake, B.E. (1968). A catalog of analyzed calciferous and sub-calciferous amphiboles together with their nomenclature and associated minerals, *Geological Society of America Special Paper*, 98p.

Leake, B.E. (1971). On aluminous and edenitic hornblendes, *Mineralogical Magazine* Vol. 38, 389-407

Leake, B.E. (1978). Nomenclature of amphiboles, *Canadian Mineralogist*, Vol. 16, pp. 501-520

Linares, J.; Regnard, J.R. & Van Dang, N. (1983). Magnetic behaviour of grunerite from Mössbauer spectroscopy, *Journal of Magnetism and Magnetic Materials*, Vol. 31-34, Pt. 2, pp. 715-716

Meerwall, E.V. (1975). A least-square spectral curve fitting routine for strongly overlapping Lorentzians or Gaussians, *Computer Physics Communications*, Vol. 9, pp. 117-128

Mitra, S. (1992). *Applied Mössbauer Spectroscopy: Theory and Practice for Geochemists and Archeologists*, Pergamon Press, Oxford, 400p.

Newton, R.C. & Perkins, D. (1982). Thermodynamic calibration of geobarometers based on the assemblages garnet–plagioclase–orthopyroxene (clinopyroxene)–quartz, *American Mineralogist*, Vol. 67, 203-222

Papike, J.J. (1988). Chemistry of the rock-forming silicates: Multiple-chain, sheet and framework structures, *Review in Geophysics*, Vol. 26, pp. 407-444

Papike, J.J.; Ross, M. & Clark, J.R. (1969). Crystal chemical characterization of clino-amphiboles based on five new structure refinements, *Mineralogical Society of America Special Paper* No. 2, pp. 117-136

Plyusnina, L. (1982). Geothermometry and geobarometry of plagioclase-hornblende bearing assemblages, *Contribution to Mineralogy and Petrology*, Vol. 80, pp. 140-146

Schumacher, R. (1991). Compositions and phase relations of calcic amphiboles in epidote and clinopyroxene bearing rocks of the amphibolite and lower granulite facies, central Massachusetts, USA, *Contribution to Mineralogy and Petrology*, Vol. 108, pp. 196-211

Seifert, F., & Virgo, D. (1973). Temperature dependence of intracrystalline Fe^{2+}–Mg distribution in a natural anthophyllite, *Carnegie Institution of Washington Yearbook* Vol. 73, pp. 405-411

Shido, F. & Miyashiro, A. (1959). Hornblende of basic metamorphic rocks, *Tokyo University Faculty of Science Journal Section* 2, Vol. 12, pp. 85-102

Skogby, H. (1987). Kinetics of intracrystalline order-disorder reactions in tremolite. *Physics and Chemistry of Minerals*, Vol. 14, pp. 521-526

Skogby, H. & Annersten, H. (1985). Temperature dependence Fe–Mg cation distribution in actinolite–tremolite, *Neues Jahrbuch für Mineralogie Monatshefte*, 193-203

Subramanium, A.P. (1956). Mineralogy and petrology of the Sittampundi complex, Salem district, Madras state, India, *Bulletin Geological Society of America*, Vol. 67, pp. 317-390

Microstructure – Hydro-Mechanical Property Relationship in Clay Dominant Soils

J. Gallier[1], P. Dudoignon[1] and J.-M. Hillaireau[2]
[1]HydrASA Laboratory, ENSIP, Poitiers University,
[2]INRA Domaine Expérimental,
France

1. Introduction

The hydro-mechanical properties of clay dominant soils are mainly governed by their clay mineral properties: i.e. mineralogy, interlayer charge, nature of exchangeable cations and associated swelling properties. The layer thickness of a clay mineral is usually measured by XRay Diffraction. Its responses to successive dry, hydration states and/or organic molecular saturation constitute the main tool for their identification (Brindley and Brown, 1980). The very great variety of these minerals is based on pure non swelling poles as kaolinite-serpentine, pyrophyllite-talc, illite–micas, chlorites and pure swelling poles as smectites. Besides these pure "end members" a lot of mixed layers as illite-smectite (I/S), kaolinite-smectite (K/S), chlorite-smectite (C/S) and many others can exist (Meunier, 2003; Brindley & Brown, 1980). These differences in mineralogy can induce differences in the hydro-mechanical properties of the clay matrices face to the shrinkage/swelling phenomenon (Tessier & Pedro, 1984; Tessier *et al.*, 1992). Consequently to the textural characteristics of the soil, they induce differences in the geotechnical properties from the weak plastic to very plastic and "liquid" domains according to the Atterberg classification. The relationships which can exist between the clay mineralogy and the hydro-mechanical properties of a soil do not depend directly of the mineralogical characteristics but mainly of the induced microstructure behaviour of these micro-divided clay media when submit to hydric or/and mechanical stress. Nevertheless, a low amount of smectite in a clay matrix assemblage can be sufficient to induce high swelling-shrinkage properties (Bernard, 2006; Bernard *et al.*, 2007). Biarez *et al.* (1987) demonstrated the similitude of the state path of a kaolinitic matrix during compression by oedometer test and during desiccation: the suction pressures induced during the desiccation cycle may be compared to the compression pressures applied during the compressibility test. The clay matrix microstructure behaviours may be represented along their shrinkage curve in volume/ water content (V-W) diagram or in "normalized" void ratio – water content (e-W) diagram. At a macroscopic scale the pedologists commonly reconstruct the soil shrinkage curves by rehydration of unremoulded but dried soil samples (Braudeau *et al.*, 1999; 2004). The method allows a characterisation of the macro-porosity of the sampled soils for accurate depths. On another way, the shrinkage curve can be obtained on the clay matrix in order to characterize the micro-arrangement of clay particles (Bernard, 2006; Bernard *et al.*, 2007).

In fact hydro-mechanical properties of a clay dominant material have to be studied at successive scales from the crystallite size to the macroscopic scale. Thus, the characterization from the microstructure-to-macroscopic scale is one of the main topic face to the explanation and model of the clay dominant material behaviours in civil engineering, pedology and soil farming domains. Many tools are currently used to measure the hydro-mechanical properties of soils. They can be *"in situ"* investigations as penetrometer, scissometer and pressiometer for mechanical parameters and "infiltration" tests for hydraulic conductivity. They can also be laboratory measurements by scissometer, triaxial cell or oedometer. All these techniques of investigation give whole characteristics of the material at the macroscopic scale.

The topic of this chapter is to propose a methodology for macroscopic-to-microscopic scale switch. The switch method is based on the use of the clay matrix shrinkage curve as tool for:

1. the calculation of numeric relationships between macroscopic *"in situ"* or laboratory measurements and the clay matrix microstructure (Perdok *et al.*, 2002; Bernard *et al.*, 2007; Bernard-Ubertosi *et al.*, 2009).
2. the modelling of the shrinkage-swelling phenomenon in the structural evolution of soils (Gallier, 2011).
3. The calculation of numeric relationship between the soil resistivity and the microstructure.

The method was induced by previous works on the clay matrix microstructure behaviours based on image analyses of thin sections captured by optical microscopy and by SEM (Dudoignon & Pantet, 1998; Dudoignon *et al.*, 2004; Dudoignon *et al.*, 2007).

The investigations were made on clay-rich soils of marsh (Atlantic coast of France). The *"in situ"* investigations consist in parallel profiles of cone resistance (Qd; dynamic penetrometer), shear strength (C; scissometer), soil resistivity (ρs; salinometer). They are coupled with water content (W) and density profiles plus 1/5 soil conductivity profiles ($CE_{1/5}$) measured on soil samples (Pons, 1997; Pons & Gerbeau, 2005).

2. Material and methods

Face to the dependence of the hydro-mechanical properties of soils on their texture and mineralogy, the studies, focussed on the mechanisms and model of clay dominant soil behaviours, require investigations on textural and mineralogical homogeneous clay material. For that, the experimental sites studied are located in the Marsh of Rochefort which belongs to the West Marsh located along the Atlantic coast of France (Figure 1a). The material consists in clay-rich soils formed by desiccation and consolidation of fluviomarine sediments which have shown:

1. their textural and mineralogical homogeneity all over the West Marsh area (Ducloux, 1989; Righi *et al.*, 1995; Pons & Gerbaud, 2005; Bernard, 2006),
2. and structural profiles which show the clay-rich material from its solid state near the surface down to plastic and liquid state in depth (Bernard, 2006. Bernard *et al.*, 2007; Bernard-Ubertosi *et al.*, 2009; Dudoignon *et al.*,2007 ; 2009 ; Figure 1b).

2.1 Geological setting

The "Marais Poitevin", the "Marais Breton-Vendéen" and the Marsh of Rochefort are the three largest marshes of the French Atlantic Coast. For this work the "*in situ*" investigations were performed in the "Marais Poitevin" and then focused on the experimental site of the INRA of St Laurent de la Prée in the Marsh of Rochefort (Figure 1a; Bernard, 2006; Gallier, 2011). The soils result from the surface desiccation and consolidation of the clay-rich fluviomarine sediments (Bri) dated of the Flandrian transgression.

The age of sediments ranges from 10 000 years BP to the present. The soil formation results mainly from the reclaiming of land from the sea by polders since the Middle Ages and consequently from compaction and maturation of salt-marsh mud. The Bri is characterized by a fine-grained texture (85-to-92% of particles < 20 µm) and small organic matter content (0.4 to 2.4%). The Cationic Exchange Capacity of the material range between 20 and 30 meq/100g. These are C.E.C. according to the "illite domain". In fact the X Ray diffractions identify the Bri clay matrix as an assemblage of dominant illite plus kaolinite and illite/smectite mixed layers and very small amount of pure smectite. The micro-chemical analyses of the assemblage spread from the illite domain to I/S domain in the M^+, $4Si$, $3R^{2+}$ Meunier's triangle (Bernard, 2006; Dudoignon *et al.*, 2009). The mineralogy is homogeneous all over the different studied sites. One can remarks only a weak increase of the smectitic layers % of the I/S with depth in the < 0.2 µm fraction. These minor mineralogical evolutions cannot imply realistic changes in the hydro-mechanical behaviour of the Bri (Bernard, 2006). The shrinkage, plasticity and liquidity limits are 20%, 40% and 70% in weight % respectively.

The structural profiles of the soil are governed by the desiccation mechanism which operates from the surface to the depth. The consequence is an increase of the water contents from the surface to the depth during the dried seasons. The hydric state of the material evolves from the shrinkage limit in surface down to the plasticity limit and liquidity limit in depth (Figures 1 and 2). The hydric state of the clay material is characterized by its water content (W), its associated wet density (γw) and its void ratio (e). Thus the structural profiles may be represented in the e-W diagram where they superimposed on the shrinkage curve of the clay matrix constituting the Bri (Figure 2b). The hydraulic management of the marshes area divides the territory in dried marsh locate in the central part of the territory and wet marshes located along the peripheral limestone – Bri contact. The role of the wet marsh is the storage of fresh water during winter.

2.2 Methods

The topic of the work is to record in parallel vertical profiles of physical and mechanical characteristics: water content (W), wet density (γw), cone resistance (Qd), shear strength (C), 1/5 soil electrical conductivity ($CE_{1/5}$) and "*in situ*" soil resistivity (ρs). The physical parameters (W, γw, $CE_{1/5}$) were measured from the surface to 2.00 m depth on samples cored using a clay-auger, the mechanical parameter (Qd, C), and soil resistivity (ρs) were obtained by "*in situ*" investigations.

The water content is calculated in weight % by difference between the "*in situ*" wet sample mass and 105°C (24 hours) dried mass referring to the dried mass. The density is measured by double weighing after paraffin coating. The porosity (n) and void ratio (e) are calculated as follows:

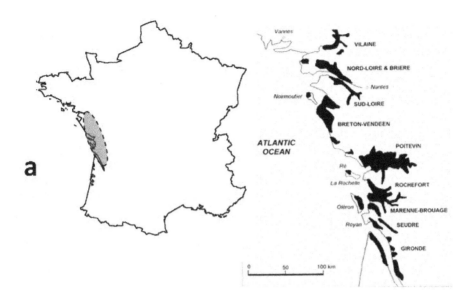

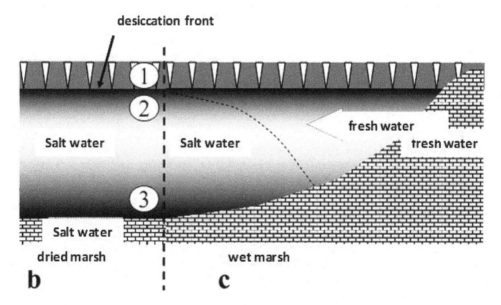

Fig. 1. a- Location of the West marshes and Mash of Rochefort along the West altlantic coast of France. b & c schematic representation of the sediment-to-soil structure evolution overlapping the limestone.(1 desiccated and consolidated soil in solid state; 2 – plastic state, 3 – liquid state).b - The Bri is saturated by fossil salt water in the dried marsh. c - It shows fresh water inlet from the peripheral limestone hills in the wet marsh.

$$n = \frac{Vv}{Vt} = \frac{Vt - \frac{Md}{\delta s}}{Vt} \tag{1}$$

and

$$e = Vv/Vs = \frac{n}{n-1} \tag{2}$$

where V_v (cm³) is the void volume (air + water), V_t is the total sample volume, Vs the solid volume, Md is the dried sample mass (g), γs is the average mineral density measured using pycnometer. It is equal to 2.58 g/cm³.

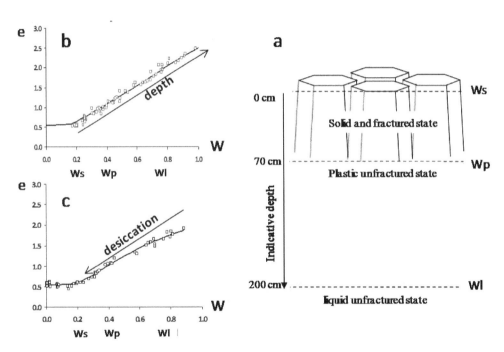

Fig. 2. The structural evolution of the clay dominant soil may be represented by the (e-W) shrinkage curve of the clay matrix, a – schematic representation of the structural profile, b-representation of the W profiles in the e-W diagram, c - clay matrix shrinkage curve obtained by drying of initial unremoulded samples. Ws = shrinkage limit, Wp = plasticity limit, Wl = liquidity limit, e = void ratio, W = gravimetric water content

The cone resistances were measured using a PANDA dynamic penetrometer with variable energy. It is a light apparatus particularly well adapted for these wetlands (Bernard et al., 2007; Bernard-Ubertosi et al., 2009). The method and its relations to other "in situ" measurements were described by Gourvés & Barjot (1995) and Langton (1999). The cone is driven into the soil using a fixed-weight hammer and the dynamic cone resistance is calculated by a microprocessor for each penetration (< 4 cm) following the so-called "Dutch Formula" (Cassan, 1988; Zhou, 1997):

$$Qd = \frac{MV^2}{2Ad} * \frac{1}{1+\frac{P}{M}} \tag{3}$$

with d the penetration (cm), M the weight of the striking mass (kg), P is weight of the struck mass (rod +cone; kg), V is the impact velocity (cm^{-1}), Qd is the cone resistance (MPa), and A is the cone section (cm^2). In these marshlands, lost cones of 4 cm^2 section (2 cm^2 rod section) have been used in order to avoid the artifact of rod-clay contact.

The shear strength was measured using a GEONOR H-70 field vane shear test. Three sizes of four-bladed vanes are used for a total range of 0-260 kPa: 16 mm (total width of two opposite blades) * 32 mm (blade height), 20*40 mm and 25.4*50.8 mm. The precision of the measurement is ± 1 kPa.

The hydraulic conductivity is measured on test pieces of Bri during oedometer compressibility tests. The experiments were performed on unremoulded core-samples around 2 m depth in saturated state near the liquidity limit. The successive steps of consolidation allow successive measurements of the hydraulic conductivity for a decrease of the void ratio from 2 (equivalent to Wl) to 0.5 (equivalent to Ws; Figures 3, 13).

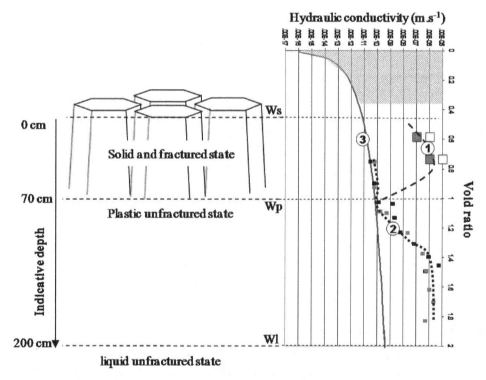

Fig. 3. Representation of the vertical evolution of the Bri structure and soil hydraulic conductivity; 1 (large squares) *"in situ"* infiltrometry measurement in the fractured surface layer (unsaturated solid state), 2 (small squares) oedometer measurement on saturated and unfractured clay material, 3 (continuous line) theoretical profile of hydraulic conductivity based on the microstructural parameters. The dashed domain does not exist *"in situ"*.

The microstructure – hydro-mechanical properties relationships were demonstrated by the petrographic studies of consolidated and sheared kaolinitic test pieces using a triaxial press (Dudoignon et al., 2004). The state ways of the simple consolidated or over-consolidated and sheared clay matrix have been represented in a Cam-clay type diagram (e versus log P). In the same time, the hydraulic conductivities of the damaged clay matrices were calculated using a Kozeny-Carman equation which took into account the micro-structure parameters as porosity, surface area of clay particles and tortuosity of the matrix (Grolier et al., 1991; Dudoignon et al., 2004):

$$K = \left(\frac{1}{2T^2}\right)\left(\frac{n^3}{Ssp^2}\right) \tag{4}$$

with Ssp= S (1-n) with S = particle surface area = specific surface / volume, T = tortuosity, n = porosity.

For the Bri, the clay matrix hydraulic conductivities are calculated using the specific surface of the clay particles (60 m²/g) calculated from the laser granulometry. The porosity (n) is calculated for each equivalent void ratio (e). To simplify, the tortuosity was calculated for average 2-5 shape factor of particles and for an isotropic medium. In the 0.5 – 2.0 void ratio range, the tortuosity evolves from 1.28 to 1.80 (Dudoignon et al., 2004; 2009).

3. Results

3.1 Gravimetric water content profiles and clay matrix shrinkage curve

The switch from gravimetric water profile to shrinkage curve is based on the calculation of the void ratio (e) for each W measurement. The e calculation needs the measurement of the sample wet density (γw) associated to each W measurement. The void ratio is calculated as follows:

$$e = \frac{\gamma s - \gamma d}{\gamma d} \tag{5}$$

with:

$$\gamma d = \gamma w/(1 + W) \tag{6}$$

γs = particle density, γw = wet sample density, γd = dried sample density, W=gravimetric water content, e= void ratio.

The presented results were obtained on the INRA experimental site of St Laurent de la Prée on grassland and cultivated field. In these marshlands, whatever the investigation site, the general soil profile shows an increase of the water content from the surface to the depth and consequently associated decrease of γw and γd (Figures 2 & 4).

The shrinkage curve can be calculated from these "in situ" data: i.e. e calculation from W and γw measurements. It shows the soil structure evolution in the Ws – Wl domain which is observed "in situ". From Wl to Ws the curve shows the void ratio decrease from 2.0 down to 0.55, and for W < Ws the void ratio is constant (e = 0.55). The shrinkage curve can be approached using the Cornelis equation (Cornelis et al., 2006; Figure 2b):

$$e = e_0 + \gamma(\exp(-\frac{\xi}{\vartheta^{\varsigma}}) \tag{7}$$

with $e_0 = 0.55$, $\gamma = 11.59$, $\xi = 1.78$, $9=W$, $\zeta=0.63$.

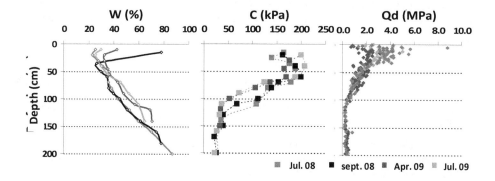

Fig. 4. Example of water content (W), shear strength (C) and cone resistance (Qd) profiles recorded in the grassland in July 2008, September 2008, April 2009 and July 2010.

The shrinkage curve has been also obtained in laboratory by drying unremoulded samples of Bri at initial water contents near the liquidity limit. The shrinkage curve has been constructed step by step from daily density and gravimetric water measurements during the drying phases up to the full desiccation at 105°C (Bernard *et al.*, 2007). This "laboratory" curve differs from the *"in situ"* one by a weak curvature in the 0.5 – 0.9 W domain (Figure 2b & c). Nevertheless, it can be modelled by a second Cornelis equation with $e_0 = 0.55$, $\gamma = 2.47$, $\xi = 0.54$, and $\zeta=1.23$. From W= 0 to W=0.50 these two shrinkage curves are perfectly superimposed. Bernard *et al.* (2007) demonstrated that the shrinkage is isotropic in the Ws - Wp domain. In fact, all the structural profiles recorded in the marsh territories, are well superimposed on the *"in situ"* shrinkage curve. The slope of the shrinkage curve is equal to the average particle density (2.58 g/cm³) measured by pycnometer (Bernard, 2006; Gallier, 2011).

This *"in situ"* shrinkage curve of the clay matrix is used as reference in the following works. It has the role of tool to establish the structure-to-hydro-mechanic relationship for these marsh soils.

3.2 Mechanical resistance profiles

3.2.1 Cone resistance and shear stress profiles

The *"in situ"* penetrometer profiles were driven down to 2 m depth in grassland (Figure 4) and in the cultivated field (Figure 5). The Qd are measured by steps of 1-to-4 cm. From the surface to the depth, the Qd profiles cross through (1) the clay material in solid state, for depths equivalent to the Ws-Wp domain, (2) the clay material in plastic state (Wp – Wl domain), and (3) the clay material in liquid state for W > Wl (Figures 4 & 5).

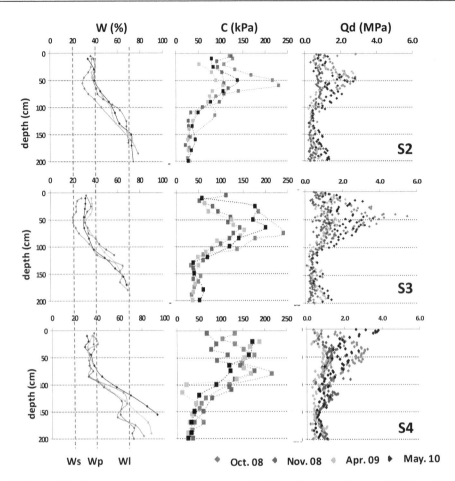

Fig. 5. Example of water content (W), shear strength (C) and cone resistance (Qd) profiles recorded in the S2, S3 and S4 sites of the corn-field in October, November 2008, April 2009 and May 2010.

The cone resistance (Qd) depends of the structure of the clay material and of its consolidation state. Thus, in the objective to get soil structure-mechanical properties relationships, the idea was to measure the Qd profiles parallel to the desiccation profiles. In fact, the successive investigations shown that the Qd profiles are able to record the desiccation effects but also superimposed structural evolutions of surface due to the tillage or other farming works plus structural evolutions in depth due to the soil over-consolidation in paleosol levels (Bernard et al., 2007, Gaillier, 2011). In order to get numerical W-e-Qd relationship with realistic correlation factors the following work was focused on soil profiles exempt of tillage zone or paleosol. The Qd profiles show vertical evolutions which are symmetric to the W profiles. According to the weak W values of surface, they present a Qd peak from the surface to the depth of 50 cm. The Qd peak maximum increases according to the desiccation intensity. From 50 cm to 100-120 cm they show a progressive

decrease of the Qd values until the inferior limit of measurement. From 100 cm to 200 cm the Qd values are very weak and constant whatever the season. These three vertical domains accord with the superimposition of the solid state (0-50 cm), solid-to-plastic state (50-100 cm) and plastic-to-liquid state (100 cm – 200 cm) (Figure 4 & 5).

The *"in situ"* shear strength (C) profiles where also driven down to 200 cm deep but by steps of 10 cm according to the clay auger sampling.

The main differences between the Qd and C measurement are due to the soil-tool interactions (Figure 6):

- a volumetric deformation and compression of the soil around the penetrometer cone
- a shearing of soil along a cylindrical surface during the shear test
- a very short time of soil damaging (few milliseconds) due to the hammer impact of the penetrometer
- a slow rotation of the shear vane for the shearing test.

Nevertheless, the vertical evolutions of the C profiles are similar to the Qd profiles: high C values for high desiccation of surface and low values for depths > 100 cm (Figures 4 & 5).

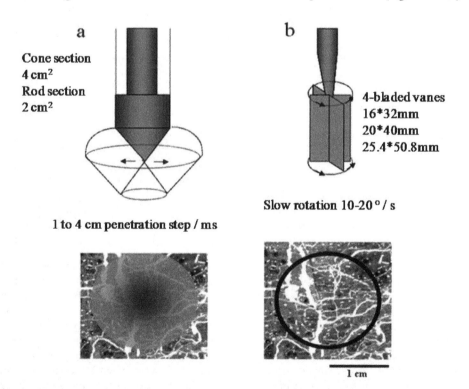

Fig. 6. Schematic representation of the impacts of the penetrometer cone and shear vanes on the soil structure. Soil photographies are grey level image of polished section. White = macroporosity and shrinkage fractures, grey = clay matrix.

3.2.2 Cone resistance and shear strength profiles – Structural profile relationship

The structural evolution of the soil from the surface to the depth may be represented by the clay matrix shrinkage curve. Thus the C and Qd profiles – W profiles relationship may be represented in a W-e-Qd or C diagram. In the e-Qd and e-C diagram the Qd and C profiles describe an hyperbolic like curve with two asymptotic ends: drastic increase of Qd an C values for low e values (desiccated and consolidated surface layer) and very low values in the plastic-to-liquid domain (Figure 7).

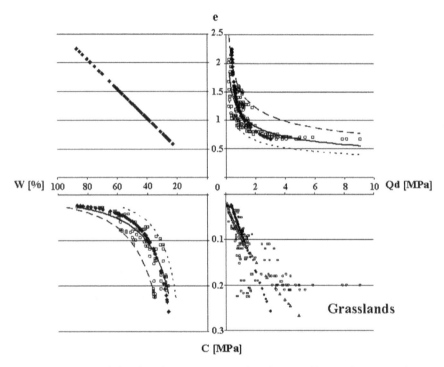

Fig. 7. Representation of the shrinkage curve – Qd and C profiles in the W-e-Qd-C crossed diagram. Example of the grassland (black line = power law equation; black lozenges = Perdok's like equation, dashed lines = limits of the domain by power law equations).

Two types of equation can model the Qd and C profiles taking into account the W or e values:

- the Perdok modified equation (Perdok *et al.* 2002; Bernard-Ubertosi *et al.*, 2009; Gallier, 2011)
- and power law equation (Gallier, 2011)

The Perdok equation is written as follows (for Qd or C):

$$\log (Qd) = a_0 + a_1 \frac{e}{1+e} + W \left(a_2 + a_3 \frac{e}{1+e}\right) \tag{8}$$

$$\log (C) = a'_0 + a'_1 \frac{e}{1+e} + W \left(a'_2 + a'_3 \frac{e}{1+e}\right) \tag{9}$$

The a_0, a_1, a_2 and a_3 Perdok's coefficients have to be fitted to minimize the sum of absolute values of the differences between measured and calculated data.

The power law is as follows (for Qd or C):

$$Qd = (e/A)^{1/b} \text{ or } Qd = (\gamma sW/A)^{1/b} \tag{10}$$

$$C = (e/A')^{1/b'} \text{ or } C = (\gamma sW/A')^{1/b'} \tag{11}$$

with A and b calculated to minimize the sum of absolute values of the differences between measured and calculated data.

The power law equation is easy to use which A and b coefficients which present weak variations between the successive runs and with A an b roles which are clearly identified on the vertical shift of the curve and on the curve shape. The Perdok's equation present some drawbacks due to the disparities between the a_0, a_1, a_2 and a_3 coefficients for different runs and lack of understanding on the roles of each parameter on the curve location and shape (Table 1).

	Qd profiles					
	Perdok's equation				Power law	
	a_0	a_1	a_2	a_3	A	b
grassland	2.89	-5.56	-0.76	1.39	1.1	-0.29
Bernard (2006)	2.34	-5.20	0.74	0.34		
cornfield	-0.35	5.68	-9.28	6.24	1.14	-0.42
	C profiles					
	Perdok's equation				Power law	
	a'_0	a'_1	a'_2	a'_3	A'	b'
grassland	-3.36	8.08	1.37	-10.39	0.44	-0.40
cornfield	1.64	-4.93	-2.08	3.35	0.36	-0.48

Table 1. Recapitulative table of Perdok's a_0, a_1, a_2, a_3 parameters and A and b power law parameters for the Qd and C profiles.

3.3 "In situ" resistivity profiles and soil electrical conductivity

3.3.1 "In situ" resistivity

The *"in situ"* soil resistivity was measured following two methods:

-   vertical resistivity (ρs) profiles using a penetrometer-salinometer coupling (Bernard-Ubertosi et al., 2009)
-   and vertical resistivity sections (Bernard, 2006; Gallier, 2011).

The penetrometer-salinometer coupling consists in driving the salinometer in the penetrometer hole after the Qd measurements. The method has the advantage to give real resistivities at measured depths (Bernard-Ubertosi et al., 2009). It has the drawback to be limited in depth. The used resistivimeter is a Syscal R1+. The salinometer device is a A-M-N-B Wenner type with 3 cm AM, MN and NB apart.

The resistivity sections are recorded following Wenner-Schlumbeger configurations. They are inverse calculated according to the apparent measured resistivity and need to be standardized. The used resistivimeter is a Syscal R1+ interfaced with a 48 electrodes switch. The measurement sequences are loaded via the ELECTRE II software. The data are transferred to a PC with the PROSYS software. The resistivity sections are calculated with RES2DINV software (Figure 8).

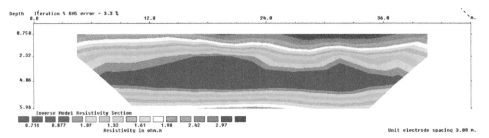

Fig. 8. Example of resistivity section showing the vertical structure of the Bri with high resistivity representative of the dried and consolidation surface level (solid state), intermediate resistivity characteristic of the plastic state and lowest resistivity around 4 meter depths for the Bri in liquid state and salt water saturated (grassland).

The parallel recording of W and ρs have allowed the fit of the Archie's law (1942) for these clay-dominant soils. Following the Archie's law the soil resistivity depends of the porosity (ϕ), saturation index (sat) and fluid resistivity (ρf):

$$\rho s = \alpha \, \rho f \, \phi^{-m} Sat^{-n} \qquad (12)$$

with α=formation factor, m = cementation factor, n = factor characteristic of the medium.

The water content – resistivity profiles relationship had hollowed the calculation of an equivalent Archie's equation for our marsh soils (Bernard-Ubertosi et al., 2009; Gallier, 2011; Figure 9):

$$\rho s = 1.01 \, \rho f \, \phi^{-2.73} Sat^{-2} \qquad (13)$$

In fact the resistivity of a clay rich soil depends of the electrolytic conductivity and the mineral surface conductivity. Thus the initial Archie's law only valid for unclayed rocks have to take into account the clay nature of minerals by their C.E.C (Waxman & Smits, 1968).

The Waxman & Smits' equation takes into account the clay mineral C.E.C as follows:

$$\rho s = \frac{Sat^n}{a \, \phi^{-m}} \left(\rho f + \frac{B \, Qv}{Sat} \right) \qquad (14)$$

with a = factor dependant of the medium, Qv = CEC, and B = parameter representative of the mobility of the exchangeable cations of the clay structures. The B parameter can be calculated as follows (Mojid & Cho, 2008):

$$B = 4.78 \, 10^{-8} \left(1 - \exp\left(\frac{-\rho f}{0.013} \right) \right) \qquad (15)$$

The Waxman & Smits calculations made with a average 25 meq/100g CEC and fluid conductivities measured on water sampled in piezometers (2 S.m^{-1} and 4 S.m^{-1} in corn field and grassland respectively) give results equivalent to the resistivities calculated following the simple Archie's law. In these salt media of coast marshes, the high fluid conductivities minimize the clay mineral surface effect.

Once more the clay matrix shrinkage curve can be used as tool for the representation of the structure-resistivity relationship in our marsh soil environment (Figure 9)

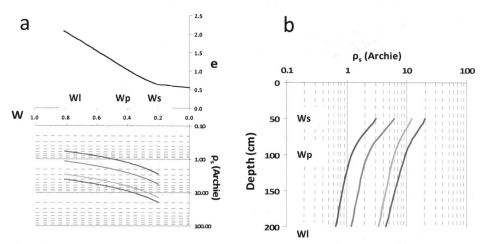

Fig. 9. Soil resistivity – shrinkage curve relationship. a- The W or e – resistivity curves are calculated in the Wr-Wp saturation domain using the Archie's law. The vertical shift of the ρs curves is due to the fluid salinity (calculations for 0.2, 0.4, 1.0, and 1.4 Ω.m). b - Equivalent vertical ρs profiles; the horizontal shift is due to the fluid salinity.

3.3.2 Electrical conductivity of soils

The *"in situ"* resistivity measurements are based on the role of porosity, saturation index and fluid resistivity. The resistivity – shrinkage curve relationship is based on the homogeneity of the fluid resistivity (Figure 9). In the marsh territories the soil properties evolve following two major mechanisms: (1) the descending progression of the desiccation fronts and (2) the fresh water – salt water exchange. The geo-electrical investigations are able to indicate the evolution of fluid salinity in soil but are not sufficiently detailed to allow realistic inverse calculation of structure –salinity-resistivity.

For this reason, profiles of 1/5 Electrical Conductivity (CE$_{1/5}$) of soils were measured. The measurements were made on soils sampled from the surface to 2.00 m depth every 10 cm, according to the clay auger sampling for water profiles. The C.E.$_{1/5}$ were measured on 1/5 extracts (10 g of dried soil in 50 g of distilled water; Pons & Gerbaud, 2005). The *"in situ"* soil electrical conductivity values depend of the water content, fluid chemical composition, soil mineralogy and the soil structure. In fact, the CE$_{1/5}$ measurements are made from 10g of dried soil, thus the CE$_{1/5}$ is "independent" from the soil water content. Nevertheless it is possible to calculate the fluid conductivity (CEf) following the Montoroi (1997) formulae:

$$CEf = CE_{1/5} (5W)$$ (16)

with CEf= the "*in situ*" fluid conductivity, CE$_{1/5}$ the soil conductivity measured in laboratory and W the gravimetric water content (Figure 10).

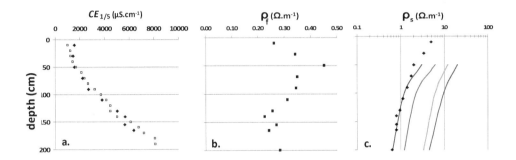

Fig. 10. Example of fluid resistivity profile (ρ_f) and soil resistivity (ρ_s) profiles calculated in the grassland from the CE$_{1/5}$ and W profiles and using the Montoroi relation plus the Archie's law. The inverse calculated profile (black points) of resistivity is well superimposed on the theoretical profile calculated with the Archie's law with a 0.2 Ω.m resistivity of salt fluid.

In the corn field the CE$_{1/5}$ profiles were recorded in September 2006, April 2007, December 2007and June 2008. For each site and whatever the date, the CE$_{1/5}$ profiles are quite superimposed (Figure 11). They are governed by the inter-seasonal water sheet level evolution and mainly by the thickness of leached soil. The CE$_{1/5}$ profile shapes evolve following the site locations:

-   Progressive evolution from the surface to the depth along the southern part of the field. This southern part is located against the limestone hill which induces fresh water inlet into the Bri.
-   Deep leached zone and drastic CE$_{1/5}$ increase in depth in the northern part of the field which is characterized by very low water level.

The CEf profiles calculated from the CE$_{1/5}$ and W profiles allow the calculation of the associated soil resistivity profiles which clearly show the vertical evolution due to the fresh plus salt water mixture along the limestone contact or due to the leaching of unsaturated upper level (Figure 12).

4. Discussion and soil structure – Hydromecanical property relationships

The mineralogical and textural "homogeneities" are characteristics of the clay dominant soils of our marshes which have been developed by polders on fluvio-marine sediments. A second advantage is the vertical evolution of the clay material from its solid state in surface

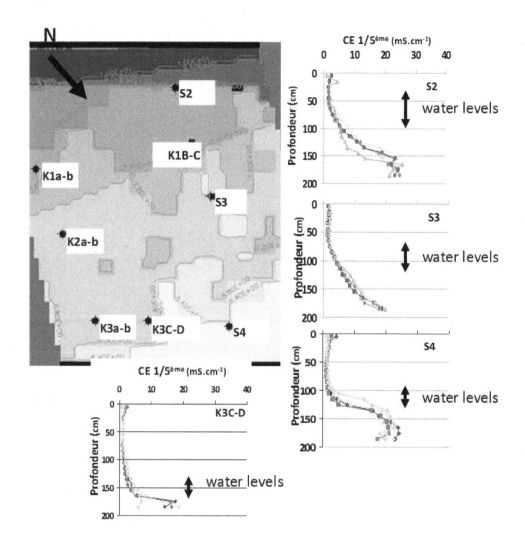

Fig. 11. Example of the $CE_{1/5}$ profile evolutions in the corn field with indication of inter-seasonal water level evolution. S2 to S4 and K1 to K3 are locations of the piezometers.

to a plastic and a liquid state in depth. In these conditions the studied marsh soils allow the transposition water content profiles– clay matrix shrinkage curve. Using the shrinkage curve the mechanical properties (Qd & C) – structure relationships may be written following power law or Perdok's like equations (Figure 13). The soil resistivity – structure relationship is an Archie's law and the hydraulic conductivity of the clay matrix may be directly linked to the shrinkage curve via experimental measurements (Figures 9, 10, 12 and 13).

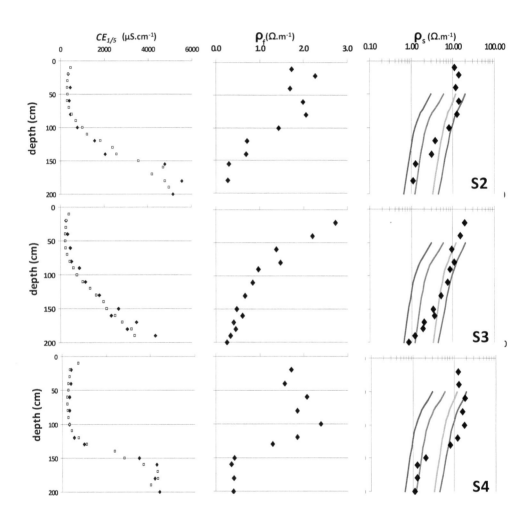

Fig. 12. Example of fluid resistivity (ρ_f) profile and soil resistivity (ρ_s) profiles calculated in the cornfield from the $CE_{1/5}$ and W profiles and using the Montoroi relation plus the Archie's law. The inverse calculated profiles show the vertical evolution of fluid salinity due to leaching in the upper unsaturated layer.

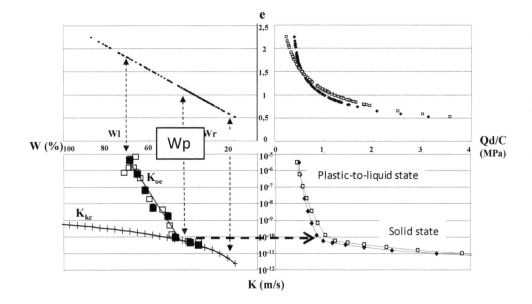

Fig. 13. Representation of the structure – hydro-mechanic properties of the marsh soils in the W-e-Qd/C-K crossed diagram. K = hydraulic conductivity, Kkc = hydraulic conductivity calculated from the clay matrix microstructure and Kozeny-Carman equation. Koe = conductivity hydraulic measured by oedometer compressivity test. The e - Qd/C curves are the average power laws characteristic of the desiccation profiles. The role of Wp is clearly shown in the Qd/C – K diagram by the differentiation of the two hydro-dynamical domains: i.e. solid state for W < Wp and plastic-to-liquid state for W>Wp.

The structure – hydro-mechanical relationship can be used to the explanation of the mechanism of soil structure behaviour (Figure 14).

These relationships based on the clay matrix shrinkage curve may be considered as dimensionless face to the hydro-mechanical behaviour of soil at the scale of the field via the geo-electrical section, at the scale of macroscopic samples from the prisms to the peds, and at the microscopic scale for particle arrangement along roots-soil contact for example (Figure 15).

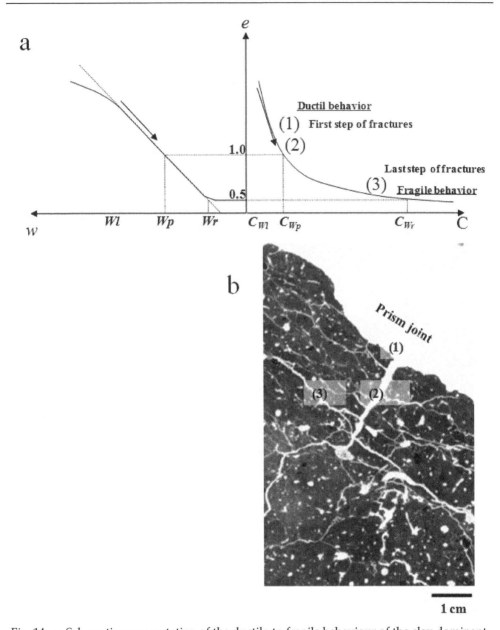

Fig. 14. a – Schematic representation of the ductile-to-fragile behaviour of the clay dominant soil taking into account the W-C couple evolution and successive steps of fracturing, b-result on the shrinkage fracture network observed on a polish section of a soil prism (grassland; from Dudoignon *et al.*, 2007 and Gallier, 2011). 1, 2 and 3 represent the successive steps of fracturing according to the shrinkage effect and increase of cohesion of the clay matrix which is associated to the desiccation effect.

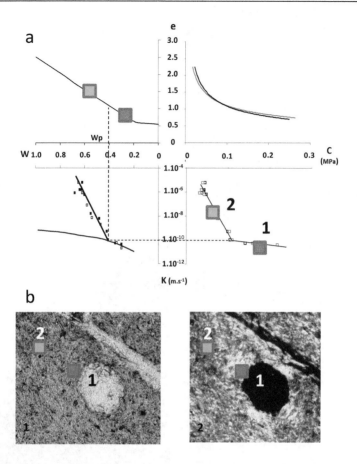

Fig. 15. Schematic representation of the hydromechanical behaviour of the soil clay matrix in a root environment. a and b - location of the 1 and 2 microsites in the root vicinity (b) and in the e-W-C-K diagram (a). 1: anisotropic and consolidated clay matrix along the root contact and 2 subisotropic and porous clay matrix 500 μm far away. The 1 and 2 microstructures of the clay matrix may be quantified by image analyses on polarized and analysed light (anisotropy) and by microanalysis on SEM (Dudoignon & Pantet, 1998; Dudoignon *et al.*, 2004).

5. Conclusion

The soil structure behaviour face to hydric of hydro-mechanic stress is governed by the facilities of the clay particles to rearrange (Biarez *et al.*, 1987; Dudoignon *et al.*, 2004). In fact the clay particle arrangement may be followed along the clay matrix shrinkage curve, thus it appears to be a good tool to make numerical relationship between the hydro-mechanical properties and structure of soil. The relationships may be described in W-e- mechanical resistance – hydraulic conductivity but also in W-e –resistivity crossed diagrams. Thus such works may be applied for modelling the soil structure behaviour for farming or geotechnic,

for modelling the shrinkage fracture network propagation introducing the cohesion in the mechanism. Finally, using the $CE_{1/5}$ and/or *"in situ"* resistivity profiles it can be aid for the modelling of hydric and salt stress in mashes territories (Gallier, 2011).

6. References

Archie, G. (1942). The electrical resistivity log as an aid in determining some reservoir characteristics. American Institute of Mining and Metallurgical Engineers, 154 pp. 1-8

Bernard, M. (2006). Etude des comportements des sols de marais: Evolution Minéralogique, Structurale et Hydromécanique. (Marai de Rochefort et Marais Poitevin). These de doctorat. Université e Poitiers. 309 p.

Bernard, M.; Dudoignon, P.; Pons, Y.; Chevallier, C. & Boulay, L. (2007). Structural characteristics of clay-dominant soils of a marsh and paleosol in a crossed diagram. European Journal of Soil Science, 58 (5) : pp. 1115-1126

Bernard-Ubertosi, M.; Dudoignon, P. & Pons, Y. (2009). Characterization of structural profiles in clay-rich marsh soils by cone resistance and resistivity measurement. Soil Science Society of America Journal, 73 (1) : pp. 46-54

Biarez, J.; Fleureau, J-M. ; Zerhouni, M-I. & Soepandji, B.S. (1987). Variations de volume des sols argileux lors de cycles de drainage – humidification. Revue Française de Géotechnique, 41 : pp. 63-71

Braudeau, E.; Costantini, J-M.; Bellier, G. & Colleuille, H. (1999). New device and method for soil shrinkage curve measurement and characterization. Soil Science Society of America Journal, 63: pp. 525-535

Braudeau, E.; Frangy, J-P. & Mohtar R.H. (2004). Characterizing non rigid aggregated soil-water medium using its shrinkage curve. Soil Science Society of America Journal, 68: pp. 359-370

Brindley, G.W. & Brown, G. (1980). Crystal structures of clay minerals and their X-ray identification. London, Mineralogical Society.

Cassan, M. (1988). Les essais in situ en mécanique des sols-réalisation et interprétation. Eyrolles, pp. 146-151

Cornelis, W.M. ; Corluy, J.; Medina, H.; Diaz, J.; Hartmann, R.; Van Meirvenne, M. & Ruiz, M.E. (2006). Measuring and modeling the soil shrinkage characteristic curve. Geoderma, 137 (1-2): 179-191

Ducloux, J. (1989). Notice explicative de la carte pédologique de France au 1/100000 Fontenay-le-Conte. 16p.

Dudoignon, P. & Pantet, A. (1998). Measurement and cartography of clay matrix orientation by image analysis and grey-level diagram decomposition. Clay Minerals, 33 : pp. 629-642

Dudoignon, P.; Gelard, D. & Sammartino, S. (2004). Cam-clay and hydraulic conductivity diagram relations in consolidated and sheared clay-matrices. Clay Minerals, 39: pp. 269-279

Dudoignon, P.; Causseque, S.; Bernard, M.; Hallaire, V. & Pons, Y. (2007). Vertical porosity profile of a clay-rich marsh soil. Catena, 70 (3): pp. 480-492

Dudoignon, P.; Bernard-Ubertosi, M. & Hillaireau, J.M. (2009). Grasslands and coastal marshes management: role of soil structure. In grasslands, Ecology, management and restore. Nova Science Publishers (NY)

Gallier, J. (2011). Caracterisation des processus d'évolution structurale et de salinité es sols de marais côtiers par mesures mécaniques et géo-électriques in situ. Thèse de doctorat, Université de Poitiers, 218 p.

Grollier,J.; Fernadez, A.; Hucher, M. & Riss, J. (1991). Les propriétés physiques des roches : théories et modèles. Masson ed., Paris, 462 p.

Gourvés, R. & Barjot, R. (1995). The PANDA dynamic penetrometer , 11th Eurprean congress of soil mechanic and foundation works, Copenhague, pp.83-88

Langton, D.D. (1999). The Panda lightweight penetrometer for soil investigation and monitoring material compaction. Ground Engineering, 32 (9): pp. 33-37

Meunier, A. (2003). Argiles. Paris 433 p.

Mojid, M.A. & Cho, H. (2008). Wetting solution on an electrical double layer contribution to bulk electrical conductivity of sand-clay mixtures. Vadose Zone Journal, 7, pp. 992-980

Montoroi, J.P. (1997). Conductivité électrique de la solution du sol et d'extraits aqueux du sol – Application à un sol sulfaté acide salé de Bassa-Casamance (Sénégal). Etude et Gestion des sols, 4 pp. 279-298

Perdok, U.D.; Kroesbergen, B. & Hoogmoed, W.B. (2002). Possibilities for modelling the effect of compression on mechanical and physical properties of various Dutch soil types. Soil and Tillage Research, 65: pp. 61-75

Pons, Y. (1997). Comportements physiques et aptitudes à la mise en culture des sols des Marais de l'Ouest (Physica behavior and soil cropping possibilities of the French West Marshes), thèse de doctorat de l'Institut National Agronomique Paris – Grignon, 130 p.

Pons, Y. & Gerbaud, A. (2005). Classification agronomique des sols de marais à partir de la relation entre sodicité et stabilité structural. Application au cas des marais de l'ouest. Etude et gestion des sols, 12 (3) : pp. 229-244

Righi, D.; Velde, B. & Meunier, A. (1995). Clay stability in clay-dominated soil systems. Clay Minerals, 30 (1) : pp. 45-54

Tessier, D. & Pédro, G. (1984). L'organisation et le comportement des sols. Association Française d'Etude des Sols, livre Jubilaire, pp. 223-234

Tessier, D. ; Lajudie, A. & Petit, J-C. (1992). Relation between the macroscopic behaviour of clays and their microstructural properties. Applied Geochemistry, Supplement 1 (Supplementary Issue 1): pp. 151-161

Waxmann, M.H. & Smits, L.J.M.(1968). Electrical conductivities in oil-bearing shaly sands. Journal of Theo Society of Petroleum Engineering, 8, pp. 107- 120

Zhou, S. (1997). Caractérisation des sols de surface à l'aide du pénétromètre dynamique léger à energie varaible type „PANDA". Thèse de Doctorat. Université Blaise Pascal, Clermont Ferrand

Mineral and Organic Matter Characterization of Density Fractions of Basalt- and Granite-Derived Soils in Montane California

C. Castanha[1], S.E. Trumbore[2] and R. Amundson[3]

[1]*Earth Sciences Division,*
Lawrence Berkeley National Laboratory, Berkeley,
[2]*Max Planck Institute for Biogeochemistry, Jena,*
[3]*Division of Ecosystem Sciences,*
University of California Berkeley,
[1,3]*USA*
[2]*Germany*

1. Introduction

There is ample evidence that soil mineralogy affects carbon cycling rates (e.g. Feller and Beare, 1997; Masiello et al., 2004; Torn et al., 1997). Humus may be physically protected from biological attack by (1) direct adhesion to clay surfaces via electrostatic interactions, hydrogen bonds, and cation bridges; (2) complexation with Fe and Al cations, amorphous oxides, and terminal atoms within the mineral structure; and (3) occlusion within mineral aggregates (Baldock and Skjemstad, 2000; Krull et al., 2003). In addition, the mineral matrix affects the stability of soil organic matter (OM) via the distribution of soil pores, water retention, O_2 diffusion, and the pH of soil water. But while methods for measuring the effects of texture, aggregation, and structure are relatively well-established (see Christensen, 1992 for a review), there is no generally accepted way of isolating discrete organo-mineral complexes, so that the effects of distinct minerals can be compared and ultimately extrapolated over a wide range of mineral and environmental conditions.

Density has routinely been used to separate soil OM fractions based on their degree of association with mineral particles (Baisden et al., 2002; Christensen, 1992; Golchin et al., 1995; Monnier et al., 1962). In addition, due to variations in the specific gravity of different minerals, it has also served as a proxy for mineral species in clay (Jaynes and Bigham, 1986; Spycher and Young, 1979) and silt (Shang and Tiessen, 1998). In this study we evaluate density as a means of separating organo-mineral complexes, and use this method to explore the role of mineralogy on OM storage and cycling.

In an early study along the western flank of California's Sierra Nevada, mafic soils were found to have higher levels of clay, carbon (C), and nitrogen (N), but lower levels of OM per unit of clay (Harradine and Jenny, 1958; Harradine, 1954). To learn more about the reasons

for these biogeochemical differences, and evaluate the direct influence of mineralogy on OM stability, we separated granitic and basaltic soils into density classes designed to isolate distinct mineral species and associated organic mater (Figure 1). In the resulting fractions, we used powder X-ray diffraction to identify the dominant mineral species, C/N and stable isotopes of C and N as indices of the degree of decomposition of the OM (Baisden et al., 2002; Ehleringer et al., 2000; Nadelhoffer and Fry, 1988), and ^{14}C measurements to infer C turnover times (Trumbore, 1993; Trumbore and Zheng, 1996). Following this detailed analysis we conducted a profile analysis of the 0-2, 2-3, and >3 g cm[-3] fractions of the basaltic soil.

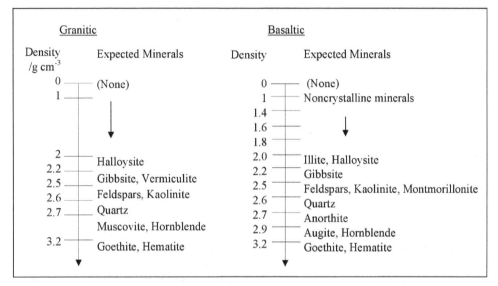

Fig. 1. Expected minerals and associated density classes for the granite and basalt soils.

2. Materials and methods

2.1 Study sites and sample selection

This study is based on three well-drained, unglaciated, forest soils: The Jimmerson, Musick, and Shaver series (Table 1). The dominant plant species are (1) ponderosa pine (*Pinus ponderosa*) and incense cedar (*Libocedrus decurrens*) at the Jimmerson site, (2) ponderosa pine, incense cedar, and manzanita (*Arctostaphylos manzanita*) at the Musick site, and (3) white fir (*Abies concolor*), sugar pine (*Pinus lambertiana*), and incense cedar at the Shaver site. The parent material of the Jimmerson soil is mapped as olivine basalt and that of the Musick and Shaver soils as granodiorite. The Jimmerson and Musick sites, with almost identical climates, are located just below the permanent winter snowline in a zone of rapid soil development, whereas the higher Shaver site is slightly cooler and wetter, and subject to a thick snowpack (Dahlgren et al., 1997). These soils represent a subset of two climate transects on which we are conducting a longer-term OM cycling study and were selected to maximise the variability in mineralogy across those transects and enable limited parent material and climate comparisons; i.e. basaltic Jimmerson versus granitic Musick and warmer Musick versus cooler Shaver.

Soil series	Jimmerson	Musick	Shaver
Parent material	Pleistocene basaltic andesite	Jurassic granodiorite	Jurassic granodiorite
Mean annual precipitation \cm	91	94	102
Mean annual temperature \°C	13.9	11.7	8.9
Elevation \m	774	1,240	1,780
Classification	fine oxidic mesic Ultic Palexeralf	fine-loamy Ultic Haploxeralf	coarse-loamy Pachic Xerumbrept
County	Shasta	Fresno	Fresno
Township & Range	NE/4 S26, T31N, R1W, MMD	S.29, T10S, R25E, MMD	S.25, T10S, R25E, MMD
Latitude & Longitude	40.52 ° N, 121.94 ° W	37.02 ° N, 119.27 ° W	37.0 3° N, 119.18 ° W

Table 1. Characteristics of the study sites (Allardice et al. 1983; Begg et al. 1985).

These soils were originally sampled, analyzed, and archived during the California Cooperative Soil-Vegetation Survey – a 47-year long reconnaissance of California's upland forests (Allardice et al., 1983; Begg et al., 1985). For the Survey, C was determined based on the mass increase of an ascarite CO_2 trap after complete combustion under a constant stream of O_2 at 900 ° C; N was determined by Kjeldahl digestion; and particle size distribution was determined by the pipette or hydrometer method.

2.2 Separation of density fractions

Density separations were performed on modern (1992) samples of the granitic soils and on archived (1961) samples of the basaltic soil. Detailed density separations were performed on the granitic Musick A1 and Shaver A2 horizons and on the A2 and Bt1 horizons of the basaltic Jimmerson soil (Table 2). Based on the specific gravity of the minerals commonly encountered in soils derived from each parent material, we isolated eight and twelve density fractions from the granite and the basalt samples, respectively (Figure 1). The <2 g cm^{-3} material of the granitic samples was separated into two fractions, but in anticipation of the possible presence of low-density allophanic minerals in the basaltic soil, the <2 g cm^{-3} material of the basaltic samples was separated into five fractions: <1, 1-1.4, 1.4-1.6, 1.6-1.8, and 1.8-2 g cm^{-3}. Following on the results of the detailed density separation, the A1 and A3 Jimmerson soil horizons were subsequently separated into just three fractions: 0-2, 2-3, and >3 g cm^{-3}.

Soil horizon samples were processed as outlined in Table 3. Successively higher density separates were obtained using a modification of the Golchin et al. (1994) method. Air-dry soils were sieved to < 2 mm, split into four ~6 g replicates, freeze-dried, and weighed into 50 ml polypropylene tubes to which ~30 ml of deionized water was added. Capped tubes were gently mixed by inverting six times and allowed to sit for at least one hour to fully wet the sample. After this period the tubes were mixed using a Vortex® mixer for ten seconds, immersed in an ice bath ice, and sonicated at 200 W for three minutes using a 350 W Branson™ sonicator with a 12.5 mm probe immersed to ~3 cm depth. Tubes were then

shaken at 180 RPM for ten minutes and centrifuged at 20,000 g to settle 1.05 g cm^{-3} and 0.2 μm diameter spherical particles. The floating material, corresponding to the 0-1 g cm^{-3} fraction, was isolated by decanting.

Series	Sampling date	Horizons	Depth\ cm	Texture	Clay\ %	Hue[a]	Structure[b]
Jimmerson	1961	A1 A2* A3 B$_t$1**	0-3 3-25 25-61 61-122	loam clay loam clay loam clay	25 33 36 51	3:1 3:1 3:1 3:1	* weak, medium, subangular blocky ** moderate, medium, angular blocky
Musick	1992	A1	0-7	loam	20	1:1	moderate, medium, granular
Shaver	1992	A2	5-10	sandy loam	7	1:1	weak, fine, granular

[a] The relative proportions of red and yellow.
[b] The strength, size, and type, respectively, of structural aggregates.

Table 2. The properties of the soil horizons examined in this study.

1. Whole air-dry soil:	Coarse split Remove big roots (~3 mm) Gently crush with mortar and pestle Sieve
2. Sieved soil:	Riffle split to obtain two replicates Add replicates to centrifuge tubes Freeze dry Weigh
3. Sequential density fractionation:	Adjust tube contents to target density Vortex Shake Sonicate Centrifuge Isolate the floating fraction and rinse it Repeat previous steps at successively higher densities
4. Density fractions:	Freeze dry Weigh Photograph Grind

Table 3. The soil processing and density separation steps.

Heavier fractions were extracted using sodium polytunsgstate solution adjusted to successively higher densities. At each step, ~30 ml of density-adjusted solution was added to each tube. Tubes were vortex-mixed to disengage the heavy fraction pellet from the bottom of the tube, shaken for ten minutes, then sonicated for 45 seconds. (The high clay

content of the Jimmerson Bt1 horizon made it extremely difficult to disengage and disperse the pellet from the bottom of the tube; for these samples, shaking and sonication times were increased as necessary). Following sonication, tubes were centrifuged to settle 0.2 μm diameter spherical particles with a density 0.05 g cm^{-3} higher than the solution. Following centrifugation the floating fractions were isolated: Floating fractions with densities less than 2 g cm^{-3} were decanted onto precombusted and tared quartz fiber filters and rinsed with 1 L deionized water. Higher density floating fractions were decanted into clean 50 ml tubes, diluted with enough water to allow them to settle, centrifuged, and rinsed 3-4 times with deionized water. All fractions were freeze-dried and weighed.

2.3 Characterization of density fractions

Morphology. Density fraction were observed and photographed through a Leica Stereo Zoom visual microscope using a Sony Cybershot digital camera.

Mineralogy. Single laboratory replicates were ground to <100 μm and placed in a Rigaku Geigerflex (Cu Kα) X-ray diffractometer. The diffraction intensity was recorded every 0.05° for 2.5 seconds by Theta software and the mineral species were identified based on the resulting powder X-ray diffraction (XRD) spectra (Barnhisel and Bertsch, 1989; Brindley and Brown, 1984).

Carbon, nitrogen, and stable isotopes. Two laboratory replicates were ground to <200 μm and analyzed on a Europa 20/20 continuous flow stable isotope ratio mass spectrometer at the Center for Stable Isotope Biogeochemistry, University of California, Berkeley. The C and N isotope ratios are reported as $\delta^{13}C$ and $\delta^{15}N$ values, where the standard is Pee Dee Belemnite carbonate for C (Kendall and Caldwell, 1998) and atmospheric N_2 for N. If replicate size was insufficient for a stable isotope measurement, replicates were either pooled or analyzed on a Carlo Erba CN Analyzer.

Radiocarbon. Single samples were weighed and sealed in evacuated Vycor tubes with 0.5g Cu, 1 g CuO, and a strip of Ag foil (Boutton, 1991), combusted for three hours at 875 ° C, then cooled (Minagawa et al., 1984). The evolved CO_2 was cryogenically purified under vacuum and measured manometrically. At the Center for Accelerator Mass Spectrometry, Lawrence Livermore National Labs, the CO_2 gas was reduced to graphite on which ^{14}C was measured and reported as $\Delta^{14}C$ (‰):

$$\Delta^{14}C = (F\text{-}1) \times 1000 \tag{1}$$

where

$$F = \frac{\left[\dfrac{^{14}C}{^{12}C}\right]_{sample(-25)}}{0.95 \times \left[\dfrac{^{14}C}{^{12}C}\right]_{1950 standard(-19)}} \times e^{\lambda(1950-y)} \tag{2}$$

F is the absolute fraction modern, the ratio between the $^{14}C/^{12}C$ of the samples, (normalized to $\delta^{13}C=$-25 ‰) and that of the international standard (95% of the activity of the NBS oxalic

acid standard in AD 1950 normalized to $\delta^{13}C=-19$ ‰). This value is corrected for the radioactive decay of the standard between 1950 and y, the year of the measurement (Stuiver and Polach, 1977). The radioactive decay constant, λ, is 1.21E-4 year^{-1}. The F values of pre-bomb atmospheric CO_2 correspond to ~1, values <1 indicate that radioactive decay to has taken place, and values >1 indicate that "bomb carbon" has been incorporated into the sample.

2.4 The carbon turnover models

A mass balance model of soil organic C states that:

$$dC_i/dt = I_i - k_iC_i \qquad (3)$$

where C_i is the carbon inventory in pool i, I is annual carbon inputs (mass year^{-1}), and k is the first-order decomposition constant (year^{-1}). Similarly, the balance of ^{14}C atoms in reservoir i, F_iC_i, can be described by:

$$d(F_iC_i)/dt = Fatm\ I - F_iC_i\ (k_i+\lambda) \qquad (4)$$

where, F_i is the ^{14}C value of pool i, FC is the ^{14}C inventory, and $Fatm$ is the ^{14}C value of the atmosphere. Starting with the common assumption that the system is in steady state with respect to ^{12}C, and hence, $I_i = k_iC_i$ (from Equation3), we used two distinct approaches to translate the ^{14}C values of density fractions into their turnover times (T_i), defined as $1/k_i$:

1. For the 1961 basalt soil fractions, which lack bomb-derived carbon, it is assumed that $Fatm$ = 1 (pre-bomb conditions). Thus, from Equation 4, $Fatm\ I = F_i\ C_i\ (k_i+\lambda)$, and

$$F_i = k_i/(k_i+\lambda) \qquad (5)$$

2. For the 1992 granite soil fractions, which contain bomb carbon, the ^{14}C value of the density fractions was translated into turnover times using a time-dependent box model. The time series was initialized in 1890, using Equation 5. In each subsequent year (t):

$$C_t F_t = I\ Fatm_{(t\text{-lag})} + C_{t-1}F_{t-1}\ (1-k-\lambda) \qquad (6)$$

where $C_t F_t$ is the ^{14}C inventory of a soil fraction in year t; $Fatm$ is the ^{14}C value of the atmosphere (Levin and Hesshaimer, 2000), and lag is the average number of years that atmospheric carbon is retained in plant tissue before becoming part of the soil OM pool. The remaining terms are defined as above. Given that at steady state, $C_t = C_{t-1}$ (and $I_i = k_iC_i$, as above), we divide equation 6 by C_t, and obtain

$$F_t = F_tatm_{(t\text{-lag})}\ k + F_{t-1}\ (1-k-\lambda) \qquad (7)$$

By matching the modelled and measured F values for the year in which the soil was sampled, the decomposition constant, k (and corresponding turnover time) can be extracted. This model assumes the fraction being modelled is homogeneous; i.e. that the decomposition rate is the same for every C atom of the population. While this assumption may be erroneous, the average turnover times derived using this approach allow for comparisons among soils and fractions.

3. Results

3.1 Morphology and mineralogy

- Whole soil

The whole soil morphological attributes (Table 2) indicate that the degree of development of the three soils increases from the Shaver, to the Musick, to the Jimmerson series. The Shaver soil had a low clay content and poorly developed structure (fragile granules 1-2 mm in diameter), whereas the Musick soil had higher clay and stronger structure (spherical granules 2-5 mm in diameter). The Jimmerson soil had the highest clay content, a distinctly redder hue (reflecting abundant iron oxides), and very strong structure (3-dimensional blocks 10-20 mm in diameter). As will become evident, these morphological differences, especially between the granite and the basalt soils, influenced the density separation results.

- Density fractions

Granitic soils. We found striking differences in the morphology and powder X-ray diffraction (XRD) patterns across the density fractions of the granite soils (Figure 2). We found a mixture of organic matter, fine roots, root bark, mineral grains, and charcoal in the 1-2 g cm^{-3} fractions; decomposed organic matter, fine roots, and kaolin clays (kaolinite and halloysite) in the 2.0-2.2 g cm^{-3} fractions; kaolins and very few roots in the 2.2-2.5 g cm^{-3} fractions; dickite (a kaolin) and feldspars (principally anorthite, but also microcline and sanidine) in the 2.5-2.6 g cm^{-3} fractions; quartz (large peak at $2\theta = 26.6°$, smaller peaks at 20.8, 50, and 59.9°) and anorthite in the 2.6-2.7 g cm^{-3} fractions; phlogopite (mica) and some anorthite or albite in the 2.7-3.2 g cm^{-3} fraction; and magnesiohornblende grains in the >3.2 g cm^{-3} fractions.

The main difference in the mineralogy of the Shaver and Musick was that, in addition to kaolins, we found gibbsite (peak at $2\theta = 18.3°$) and hydroxyl-interlayered vermiculite in the 2.2-2.5 g cm^{-3} Shaver fraction.

Basaltic soil. From 0 to 2 g cm^{-3} in the A2 and Bt1 horizons of the Jimmerson soil, OM content and sample heterogeneity decreased steadily and the OM changed from recognizable plant parts to more disintegrated and decomposed material (Figure 3a).

The diffraction patterns of the mineral density fractions, which were similar for A and B horizons, changed gradually with density (Figure 3b). Halloysite dominated the spectrum between 2.4 and 2.6 g cm^{-3}, and remained an important phase up to 2.9 g cm^{-3}. We found cristobalite in the 2.2-2.4 g cm^{-3} fraction; feldspar and quartz grains between 2.4 and 2.9 g cm^{-3} (A horizon) or between 2.6 and 2.9 g cm^{-3} (B horizon); quartz in the 2.6-2.7 g cm^{-3} fractions (its peak dwarfed all others and only the base is shown); and anorthite and/or albite in the 2.7-2.9 g cm^{-3} fraction. The orange skins we observed in the 2.2-2.7 g cm^{-3} fractions in the B horizon, which did not produce diagnostic XRD patterns, were presumably amorphous iron oxides. From 2.7 to 3.2 g cm^{-3} a transition occurred from fine halloysite particles to large reddish and metallic silver particles (A horizon) or to red/yellow particles (B horizon). Above 3.2 g cm^{-3} goethite and hematite phases dominate. The peak ratio at 35.6° versus 33.2° signifies relatively more goethite in the B horizon, which agrees with the difference in hue: More red in the A horizon and more yellow in the B horizon.

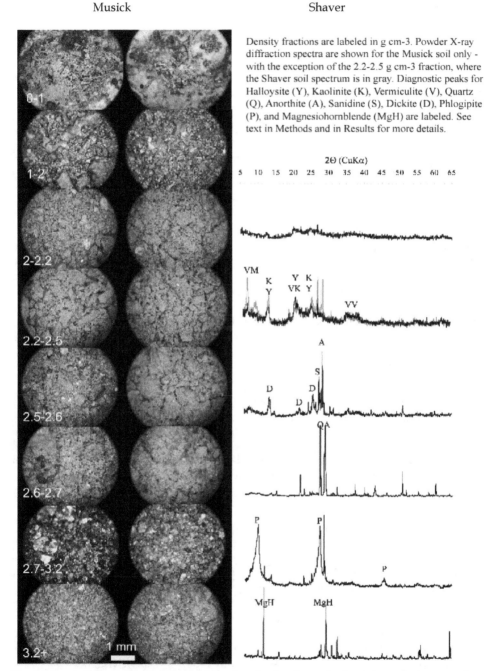

Musick

Shaver

Density fractions are labeled in g cm-3. Powder X-ray diffraction spectra are shown for the Musick soil only - with the exception of the 2.2-2.5 g cm-3 fraction, where the Shaver soil spectrum is in gray. Diagnostic peaks for Halloysite (Y), Kaolinite (K), Vermiculite (V), Quartz (Q), Anorthite (A), Sanidine (S), Dickite (D), Phlogipite (P), and Magnesiohornblende (MgH) are labeled. See text in Methods and in Results for more details.

Fig. 2. Morphology and mineralogy of the density fractions obtained from the two granite soils, Musick (left hand side) and Shaver (right hand side)

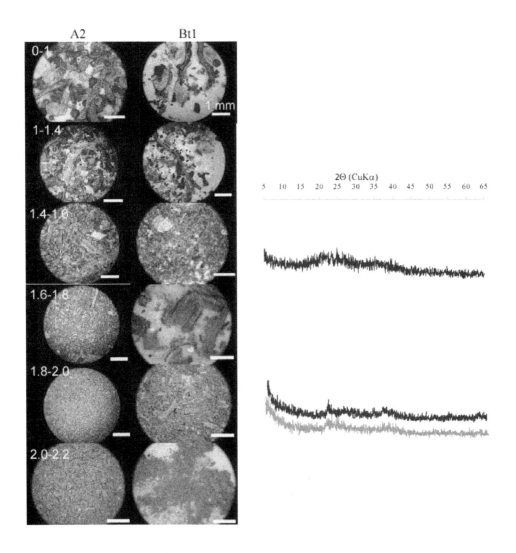

Fig. 3a. Morphology and mineralogy of the low-density mineral-free fractions obtained from the A2 and Bt1 horizons of the Jimmerson (basalt) soil. Density fractions are labelled in g cm-3. Powder X-ray diffraction spectra are shown for the A2 horizon in black and for the Bt1 horizon in gray.

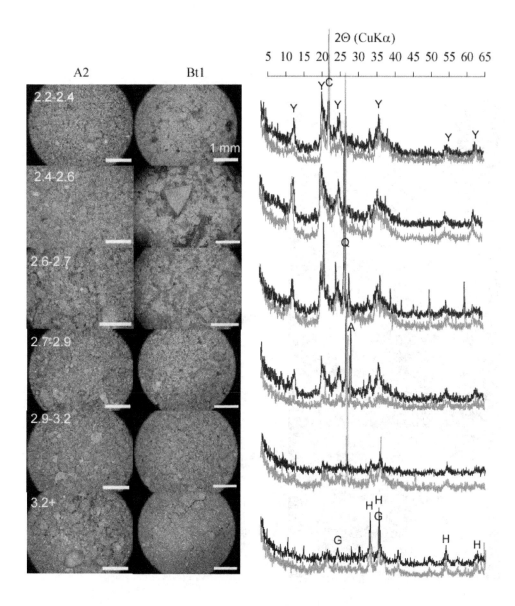

Fig. 3b. Morphology and mineralogy of the high-density mineral-associated fractions obtained from the A2 and Bt1 horizons of the Jimmerson (basalt) soil. Density fractions are labelled in g cm^{-3}. Powder X-Ray diffraction spectra are shown for the A2 horizon in black and for the Bt1 horizon in gray. Peaks for Halloysite (Y), Cristobalite (C), Quartz (Q), Anorthite (A), Goethite (G), and Hematite (H) are labeled. See text in Methods and in Results for more details.

3.2 Chemistry

- Whole soil

The % C profiles for the three soils were very similar, but the C/N and % clay profiles were quite distinct (Figure 4). There was no clear association between % C and clay values. The % clay and C/N ratios were inversely correlated, however, indicating that clay has a positive effect on the overall state of OM decay. The overall linear R-square=0.88, p<0.0001, n=10. For the Musick and Jimmerson soils (n=4 sampling depths each) the R^2 and p-values for the linear regression of C/N on % Clay are 0.99 and <0.004 for both cases. For the Shaver soil, with only two sampling depths, a regression analysis was not warranted. The whole soil $\delta^{13}C$ and $\delta^{15}N$ values increased with depth, a trend that has been observed in a number of soils (e.g. Nadlehoffer and Fry 1988).

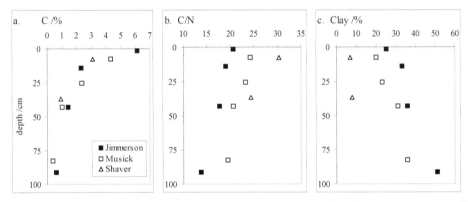

Fig. 4. Depth profiles of whole soil a. % C, b. C/N, and c. % clay for the Jimmerson, Musick, and Shaver soils (Allardice et al. 1983; Begg et al. 1985). The basalt soil is represented by closed symbols and the granite soils are represented by open symbols.

- Density fractions

The complete C, N, and isotope results of the detailed density separations are tabulated in Tables 4 and 5. Table 6 shows the results of the ANOVAs and subsequent multiple comparison tests among density fractions for each soil. To compare trends in the density fractions across sites, some of the data from tables 4 and 5 are shown in figure 5, where stable isotope values were plotted as Δ, the difference between the soil fraction δ values and the root δ values; this correction accounts for site differences in the isotopic composition of the plant inputs.

The 0 -1 g cm^{-3} density class was distinguished by high % C, low C:N ratios and high $\Delta^{15}N$ values. In the different fractions of the 1 - 2 g cm^{-3}continuum of the basaltic Jimmerson soil, % C and C/N decreased while stable isotope values generally increased (Table 6). The presence of charcoal in the 1 -1.6 g cm^{-3} density classes was reflected in C/N ratios > 40 (Table 5, Figure 5b, c). In all three soils the 2.0-2.4 g cm^{-3} density fractions corresponded to a transition between less decomposed mineral-free and more decomposed mineral-bound organic matter where % C and C/N decreased- and ^{13}C and ^{15}N increased. Higher stable isotope values and lower C/N ratios occurred at densities greater than 2.2 g cm^{-3}, a class dominated by minerals with little associated OM. The primary mineral-dominated fractions, with densities > 2.5 g cm^{-3}, were relatively unimportant with respect to C and N storage.

Density fraction /g cm^{-3} (mineral in fraction)	Mt mean	Mt s.e.	Ct mean	Ct s.e.	Nt mean	Nt s.e.	C/% mean	C/% s.e.	N/% mean	N/% s.e.	C:N mean	C:N s.e.	$\delta^{13}C$/‰ mean	$\delta^{13}C$/‰ s.e.	$\delta^{15}N$/‰ mean	$\delta^{15}N$/‰ s.e.	^{14}C F	^{14}C error	^{14}C F	^{14}C error
Musick																				
Bulk measured							4.0	0.5	0.18	0.0	22.8	2.8	-25.6	0.1	1.2	0.2	1.097	0.004		
Roots							30.1	5.8	0.71	0.1	42.4	11.0	-26.0	0.2	-2.0	0.8				
0-1 (free OM)	0.001	0.000	0.01		0.01		31.4		1.82	0.0	17.2		-26.2		1.3				1.137	0.004
1-2 (free OM)	0.063	0.001	0.60	0.07	0.47	0.02	38.7	0.0	1.31	0.0	29.6	0.8	-26.3	0.1	-0.1	0.2				
2-2.2 (Kaolin)	0.035	0.001	0.11	0.01	0.18	0.01	12.5	0.4	0.85	0.0	14.6	0.5	-25.0	0.1	2.2	0.0	1.064	0.004		
2.2-2.5 (Kaolin)	0.170	0.003	0.09	0.01	0.21	0.01	2.1	0.1	0.20	0.0	10.5	0.4	-24.2	0.0	3.2	0.0	1.068	0.004		
2.5-2.6 (Feldspars)	0.182	0.007	0.03	0.00	0.05	0.00	0.6	0.0	0.05	0.0	11.3	0.3	-24.2	0.0	2.9	0.3	1.038	0.004		
2.6-2.7 (Quartz)	0.356	0.013	0.01	0.00	0.02	0.00	0.1	0.0	0.01	0.0	9.4	0.7	-25.0	0.1	2.3	0.4	0.952	0.004		
2.7-3.2 (Micas)	0.135	0.010	0.01	0.00	0.04	0.00	0.3	0.1	0.05	0.0	7.3	1.5	-25.1	0.0	1.4	0.0			1.014	0.004
3.2+ (Hornblende)	0.055	0.006	0.00	0.00	0.01	0.00	0.2	0.0	0.02	0.0	9.9	0.6	-25.3		5.3					
Bulk recovered	0.997	0.019	0.87	0.07	0.98	0.02	3.5	0.0	0.17	0.00	19.9	0.6	-22.4	1.9	1.3	0.1	1.113	0.171		
Shaver																				
Bulk measured							1.4	0.0	0.06	0.00	22.4	1.3	-24.8	0.0	4.1	0.1	1.060	0.004		
Roots							27.8	5.5	0.58	0.05	47.3	10.2	-26.1	0.1	2.4	0.3				
0-1 (free OM)	0.002	0.000	0.11	0.02	0.06	0.01	72.5	6.6	1.70	0.18	43.4	6.1	-24.9	0.1	4.3	0.2			1.084	0.004
1-2 (free OM)	0.021	0.001	0.32	0.04	0.22	0.03	22.4	2.8	0.68	0.07	32.8	5.3	-24.1	0.5	2.2	0.2				
2-2.2 (Kaolin)	0.017	0.000	0.12	0.00	0.15	0.01	9.9	0.1	0.56	0.02	17.8	0.5	-24.5	0.1	4.8	0.1	1.039	0.004		
2.2-2.5 (Kaolin)	0.136	0.010	0.14	0.01	0.27	0.03	1.4	0.1	0.12	0.01	11.7	1.1	-23.7	0.0	5.1	0.0	1.049	0.005		
2.5-2.6 (Feldspars)	0.169	0.002	0.04	0.00	0.08	0.01	0.4	0.0	0.03	0.00	12.0	0.5	-24.0	0.3	4.8	0.1	1.045	0.004		
2.6-2.7 (Quartz)	0.414	0.004	0.03	0.00	0.05	0.00	0.1	0.0	0.01	0.00	12.5	0.6	-24.3	0.0	3.9	0.0	1.044	0.005		
2.7-3.2 (Micas)	0.173	0.002	0.02	0.00	0.07	0.00	0.2	0.0	0.03	0.00	6.8	0.3	-23.8	0.2	4.5	0.0			1.047	0.005
3.2+ (Hornblende)	0.079	0.002	0.01	0.00	0.01	0.00	0.1	0.0	0.01	0.00	12.3	0.8	-25.2	0.0	5.3	0.0				
Bulk recovered	1.010	0.011	0.80	0.05	0.91	0.04	1.1	0.1	0.06	0.00	17.7	1.7	-19.3	2.2	3.8	0.2	1.067	0.097		

Table 4. Granite soil density fraction characteristics. Proportion of total mass (Mt), proportion of total C (Ct), proportion total N (Nt), % C, % N, C/N, $\delta^{13}C$, $\delta^{15}N$, and ^{14}C (F) for the bulk soil, roots, and mineral/density fractions.

Density fraction /g cm⁻³ (mineral in fraction)	Mt mean	Mt s.e.	Ct mean	Ct s.e.	Nt mean	Nt s.e.	C/% mean	C/% s.e.	N/% mean	N/% s.e.	C:N mean	C:N s.e.	δ¹³C/‰ mean	δ¹³C/‰ s.e.	δ¹⁵N/‰ mean	δ¹⁵N/‰ s.e.	¹⁴C F	¹⁴C error
A2 horizon																		
Bulk							2.0	0.0	0.11	0.00	18.2	0.4	-25.0	0.1	6.1	0.7	0.957	0.004
Roots							38.0	1.6	1.08	0.03	35.1	9.5	27.0	0.0	-0.5	0.1		
0-1 (free OM)	0.000	0.00	0.01	0.00	0.00	0.00	43.9	1.5	1.15	0.07	38.4	2.7	-27.8	0.1	5.6	0.1		
1-1.4 (free OM)	0.003	0.00	0.07	0.04	0.02	0.01	48.8	0.7	0.64	0.03	76.9	3.3	-27.7	0.0	2.7	0.1	0.953	0.004
1.4-1.6 (free OM)	0.008	0.00	0.15	0.00	0.06	0.00	45.0	0.0	0.93	0.06	48.6	3.0	-26.0	0.0	3.1	0.0		
1.6-1.8 (free OM)	0.006	0.00	0.10	0.02	0.05	0.01	35.2	0.4	0.89	0.00	39.5	0.5	-25.7	0.1	3.4	0.1		
1.8-2.0 (free OM)	0.007	0.00	0.08	0.01	0.05	0.00	23.6	1.1	0.86	0.02	27.6	1.4	-25.8	0.0	4.5	0.0		
2.0-2.2 (Halloysite)	0.006	0.00	0.03	0.00	0.02	0.00	9.5	0.7	0.45	0.01	21.0	1.7	-25.4	0.1	5.7	0.1		
2.2-2.4 (Halloysite)	0.064	0.01	0.07	0.00	0.10	0.03	2.2	0.6	0.18	0.01	12.9	3.3	-24.6	0.2	7.1	0.2		
2.4-2.6 (Halloysite)	0.601	0.02	0.28	0.01	0.46	0.00	0.9	0.0	0.08	0.00	11.0	0.4	-23.7	0.1	8.2	0.1	0.974	0.004
2.6-2.7 (H+Quartz)	0.117	0.01	0.02	0.00	0.04	0.00	0.4	0.0	0.04	0.00	10.6	0.4	-23.8	0.1	7.7	0.1		
2.7-2.9 (H+Anorthite)	0.082	0.00	0.02	0.00	0.03	0.00	0.5	0.0	0.04	0.00	12.8	1.1	-23.9	0.0	7.6	0.0		
2.9-3.2	0.031	0.00	0.01	0.00	0.02	0.00	0.7	0.0	0.06	0.00	12.2	0.4	-23.9	0.1	7.7	0.1		
3.2+ (Iron oxides)	0.097	0.00	0.03	0.00	0.03	0.00	0.6	0.0	0.03	0.00	20.4	0.7	-23.9		7.3		0.940	0.004
Bulk recovered	1.023	0.01	0.87	0.05	0.87	0.03	1.8	0.1	0.10	0.00	15.9	1.0	-21.7	1.2	6.1	0.2	*0.965*	
Bt1 Horizon																		
Bulk							0.4	0.0	0.03	0.00	14.0	0.4	-23.4	0.1	9.2	0.0	0.879	0.003
0-1 (free OM)	0.000	0.000	-0.02	0.02	0.00	0.00	49.9	1.1	0.76	0.16	68.8	14.7	-28.4					
1-1.4 (free OM)	0.001	0.000	0.05	0.05	0.00	0.00	51.3											
1.4-1.6 (free OM)	0.001	0.000	0.06	0.02	0.00	0.00	41.6	1.5	0.45		95.8		-25.7		2.4		0.754	0.003
1.6-1.8 (free OM)	0.000	0.000	-0.03	0.01	-0.01	0.00	36.2	0.6	0.40	0.04	90.8	10.0						
1.8-2.0 (free OM)	0.001	0.000	0.05	0.00	0.01	0.00	34.0		0.64		53.3							
2.0-2.2 (Halloysite)	0.000	0.000	0.00	0.00	0.00	0.00	11.8		0.40		29.7							
2.2-2.4 (Halloysite)	0.010	0.001	0.02	0.00	0.02	0.00	0.8	0.2	0.06	0.01	12.5	3.4	-25.7		6.6			
2.4-2.6 (Halloysite)	0.679	0.079	0.52	0.02	0.69	0.10	0.3	0.0	0.03	0.00	9.4	0.4	-22.9	0.0	10.3	0.0	0.802	0.003
2.6-2.7 (H+Quartz)	0.053	0.004	0.04	0.00	0.05	0.00	0.3	0.0	0.03	0.00	11.0	0.4	-24.0	0.0	9.4	0.2		
2.7-2.9 (H+Anorthite)	0.035	0.001	0.03	0.01	0.03	0.00	0.3	0.1	0.02	0.00	13.4	3.0	-24.6	0.1	6.2	0.7		
2.9-3.2 (H+Anorthite)	0.018	0.001	0.02	0.00	0.01	0.01	0.5		0.03		17.8							
3.2+ (Iron oxides)	0.112	0.001	0.12	0.00	0.05	0.00	0.4	0.0	0.01	0.00	36.7	1.1	-23.3	0.1	8.6	0.4	0.796	0.003
Bulk recovered	0.985	0.004	0.85	0.00	0.85	0.00	0.4	0.0	0.03	0.00	9.4	1.0	-19.9	1.4	8.3	1.0	*0.797*	

Table 5. Jimmerson (basalt) A2 and Bt1 horizon density fraction characteristics. Proportion of total mass (Mt), proportion of total C (Ct), proportion total N (Nt), % C, % N, C/N, δ13C, δ15N, and 14C (F) for the bulk soil, roots, and mineral/density fractions.

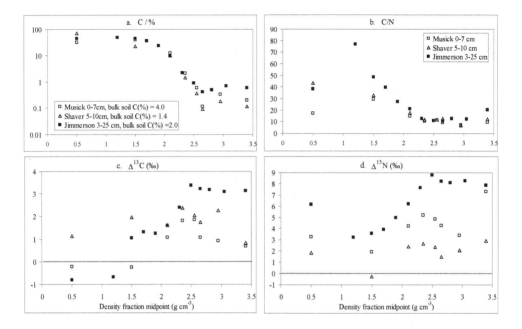

Fig. 5 a-d. The carbon and nitrogen chemistry of the A horizons plotted against the midpoint of each density fraction. The basalt soil is represented by closed symbols and the granite soils are represented by open symbols.

The proportion of total C (Ct) in each density fraction, i, is $Ct_{(ix)} = \%C_{(ix)} \times Mt_{(ix)} \div \%C_{bulk(avg)}$, where, x is the laboratory replicate for i. Reported Ct values are the mean of two laboratory replicates. Nt is calculated analogously. Where reported, the standard error (s.e.) for Mt, C, N, ^{13}C, and ^{15}N is the absolute standard error of the mean of laboratory replicates (n=4 for mass fractions and n=2 for others). Standard errors for C/N, Ct, Nt, and recovered bulk are calculated using gaussian error propagation. The error for ^{14}C is the analytical error reported by CAMS after rerunning one sample several times. The 0-1 and 1-2 g cm-3 fractions as well as the >2.5 g cm-3 were composited for ^{14}C analysis.

The granitic soils yielded density fractions that clearly differed from one another with respect to mineral and OM composition. Radiocarbon values decreased as a linear function of density, albeit with different slopes for Musick and Shaver (Table 7). The 2 g cm-3 boundary separating C cycling times of less than and greater than 100 years but differences across higher density fractions were relatively small. The only outstanding C turnover pool corresponded to the mica-dominated fraction of the Musick soil, with a ~600 year mean residence time. The 1992 granitic soil samples reflect the incorporation of ^{14}C from nuclear weapons, whereas the 1961 basalt soil samples do not. As a result we cannot directly compare these two sets of ^{14}C values and derived turnover times.

Soil series	Horizon	Density / gcm^{-3}	C / %	N / %	C:N	δ^{13}C / ‰	δ^{15}N / ‰
Jimmerson	A2	0-1	b	a	b	a	bc
		1-1.4	a	c	a	a	a
		1.4-1.6	b	b	b	b	a
		1.6-1.8	c	b	b	bc	a
		1.8-2	d	b	c	bc	abc
		2-2.2	e	d	cd	c	bc
		2.2-2.4	f	e	d	d	cd
		2.4-2.6	f	e	d	e	d
		2.6-2.7	f	e	d	e	cd
		2.7-2.9	f	e	d	e	cd
		2.9-3.2	f	e	d	e	cd
		3.2+	f	e	cd	e	cd
ANOVA P values:			<0.0001	<0.0001	<0.0001	<0.0001	<0.0001
Jimmerson	Bt1	0-1	a	a	a	-	-
		1-1.4	a	-	-	a	-
		1.4-1.6	b	ab	a	b	a
		1.6-1.8	c	ab	ab	-	-
		1.8-2	c	ab	ac	-	-
		2-2.2	d	ab	ab	-	-
		2.2-2.4	e	b	b	b	b
		2.4-2.6	e	b	b	e	c
		2.6-2.7	e	b	bc	d	c
		2.7-2.9	e	b	bc	c	b
		2.9-3.2	e	b	bc	-	-
		3.2+	e	b	bc	e	c
ANOVA P values:			<0.0001	0.0002	0.0002	<0.0001	<0.0001
Musick	A1	0-1	b	a	b	a	ab
		1-2	a	b	a	a	a
		2-2.2	c	c	b	b	bc
		2.2-2.5	d	d	c	c	c
		2.5-2.6	e	e	c	c	c
		2.6-2.7	e	e	cd	b	bc
		2.7-3.2	e	e	d	b	b
		3.2+	e	e	cd	b	d
ANOVA P values:			<0.0001	<0.0001	<0.0001	<0.0001	0.0001
Shaver	A2	0-1	a	a	a	ab	bc
		1-2	b	b	a	ab	a
		2-2.2	c	b	b	ab	cde
		2.2-2.5	c	c	b	a	de
		2.5-2.6	c	c	b	ab	cde
		2.6-2.7	c	c	b	ab	b
		2.7-3.2	c	c	b	a	bcd
		3.2+	c	c	b	b	e
ANOVA P values:			<0.0001	<0.0001	<0.0001	0.0153	<0.0001

Table 6. Results of the ANOVA and Tukey-Kramer Honestly Significant Differeces test conducted for each soil horizon and chemical analysis. P-values are for the overall ANOVA and within each of these groupings. Different letters denote significant differences among all density fractions for alpha=0.050.

In both the A2 and Bt1 horizons of the basaltic soil, high levels of clay, iron oxides, and resulting aggregation hindered the segregation of its constituent minerals. Morphology and mineralogy changed very gradually with density, such that the seven mineral density fractions yielded only two discrete organo-mineral fractions dominated by halloysite versus iron oxides. In our attempts to disperse the soil minerals, we used relatively high levels of ultrasonic energy on all samples, and this contributed to substantial losses of dissolved C (which, by difference, is ^{13}C-depleted) into the polytungstate density solution (Tables 4, 5). Most of the recovered C and N corresponds to the mineral-free (<2 g cm^{-3}) and kaolin-bearing (2-2.6 g cm^{-3}) fractions.

Density fraction / g cm-3	Radiocarbon values / FM / years	
	Musick	Shaver
0-2	1.137	1.084
2-2.2	1.064	1.039
2.2-2.5	1.068	1.049
2.5-2.6	1.038	1.045
2.7-3.2	0.952	1.044

Table 7. Radiocarbon values of the granitic soil density fractions. For Musick, ^{14}C (FM) = 1.235 - 0.084 density midpoint (g cm^{-3}), with R^2=0.85, p=0.026, n=5 and for Shaver, ^{14}C (FM) = 1.099 - 0.021 density midpoint (g cm^{-3}), with R^2=0.75, p=0.059, n=5.

3.3 Jimmerson soil profile analysis

In both A2 and Bt1 horizons the large 2.4-2.6 g cm^{-3} halloysite fraction captured over 25 % of the total C; with the largest pool of remaining C in either the 1.4-1.6 g cm^{-3} mineral-free fraction (A2 horizon) or the >3 g cm^{-3} goethite/hematite fraction (Bt1 horizon). To avoid costly AMS ^{14}C measurements on fractions with minor quantities of C we limited the ^{14}C measurements to these three fractions and the bulk soil. The weighted ^{14}C value of these three fractions (free, halloysite, and iron-oxide) is not very different from that of the bulk soil, and this provides some assurance that we did not miss a substantial, distinct, C pool.

To complete the Jimmerson soil profile analysis (Figure 6) we separated the remaining A1 and A3 horizons into just three fractions, 0-2, 2-3, and >3 g cm^{-3}, corresponding to free, halloysite, and iron oxide-bound OM, respectively. The high C/N values in the mineral-free fraction of the A2, A3, and Bt1 horizons indicate the presence of charcoal. Before deriving turnover times we made a conservative adjustment of the ^{14}C values of the free fractions with C/N > 40 by assuming that the charcoal has 85 % C, 0 % N, and a ^{14}C value equal to that of the slowest cycling pool of each horizon. In the halloysite fraction, the proportion of total C, as well as its ^{14}C-derived turnover time and stable isotope value increased steadily and predictably with depth (i.e. deeper soil → slower turnover time → higher δ^{15}N and δ^{13}C). In contrast, the isotope patterns associated with the iron-rich fractions do not vary as regularly. From the A1 to the A3 horizons the δ^{15}N values were negatively correlated with turnover time (i.e. slower turnover time → less ^{15}N) and only in the B horizon did both values increase.

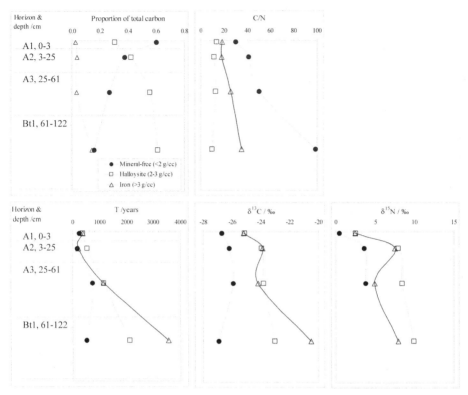

Fig. 6. Carbon inventory, C/N, ^{14}C-derived turnover time, ^{13}C, and ^{15}N in mineral-free, halloysite, and iron oxide fractions (<2, 2-3, and >3 g cm^{-3}, respectively) in the Jimmerson soil (basalt) profile. In A2 and Bt1 horizons, ^{14}C values for the 1.4-1.6, 2.4-2.6 and >3.2 g cm^{-3} fractions were presumed to be representative of the 0-2, 2-3, and >3 g cm^{-3} fractions, respectively. Before deriving turnover time we corrected ^{14}C values for charcoal based on C/N values in excess of 40 (see text in Results).

4. Discussion

The objectives of this study were to use density to separate discrete organo-mineral complexes from both granite and basalt-derived soils so as to evaluate the direct effect of parent material, and resulting soil mineralogy, on OM dynamics. Our results allow us to compare (1) the chemistry and turnover time of different density fractions within a given soil sample and (2) the chemistry of A horizons from soil with differing parent material and/or climate. In addition, (3) we examined depth related trends in chemistry and turnover of the basalt soil profile.

4.1 Evaluation of the effectiveness of the density separation method

The absence of broad bands around $2\theta = 5.9$, 26.2, and 35.9 ° confirmed that the basaltic soil did not contain allophane or imogolite phases; its low density fractions were all

mineral-free. The frequently high $\Delta^{15}N$ and low C/N values of the 0-1 g cm^{-3} are probably associated with the increased presence of ectomycorrhizal fungal products (Hogberg, 1997), such as spores and hyphae corresponding to the black spheres and spongy material visible in figures 2 and 3a. The entire 1-2.4 g cm^{-3} interval of both soil types represented a continuum of alteration from fresh litter, with high % C, N, and C/N, and low $\Delta^{13}C$ and ^{15}N, to highly humified kaolin-associated OM at the other end of the spectrum (Ehleringer et al., 2000; Nadelhoffer and Fry, 1988). This result, which was true for both the basalt and the granite samples, is consistent with both Golchin et al. (1995) and Baisden et al. (2002), and suggests that the quantity of OM associated with the mineral surface is inversely related to its degree of decomposition. The corresponding ^{14}C analyses indicate, however, that neither of these two parameters is useful for predicting the mean residence time of the organic C.

Previous radiocarbon work on the granite soils has shown that acid/base hydrolysis of the >2 g cm^{-3} mineral fraction can leave a substantial ^{14}C-depleted residue (Trumbore and Zheng 1996). The only density separates obtained in this study that are large enough to account for this pool of distinctly older C are the secondary clay-dominated fractions. If some of the C in these fractions is much older, the average residence times we report actually represent a heterogenous mixture of faster and more slowly cycling C components. This condition violates the assumption of homogeneity inherent in our C cycling model (Equation 7), and the turnover times of the kaolin-rich, and possibly other, fractions should therefore be cautiously interpreted as average ^{14}C ages of mixed C pools. We conclude that acid/base hydrolysis may be the best way to separate very long-lived C from these clay fractions. But we also note that the sodium pyrophosphate used in the base hydrolysis dissolves metastable iron chelates that can bind long-lived C (McKeague et al., 1971; Trumbore and Zheng, 1996). Thus, in iron-rich samples such as the Jimmerson B horizon, the very heavy fraction (> 3 g cm^{-3}) may be appropriate for separating this potentially important pool.

4.2 Organic matter dynamics as a function of parent material and climate

Harradine and Jenny (1958) observed that basaltic soils in California have more OM than granitic soils. In this study we saw no difference in the total C % profiles of the basaltic Jimmerson and the granitic Musick soils (Figure 4) and when plotted as a function of depth, the mineral-bound C per unit clay or the stable isotope enrichment values in the clay-dominated density fractions were indistinguishable (Figure 7, 8). The observations by Harradine and Jenny (1958) may have been due to the presence of allophanes in younger basaltic soils that likely were part of their data set, since allophane minerals can store large amounts of C. In this study, the Jimmerson soil we examined is on an old relatively stable landscape, and any allophanes initially formed have weathered to more crystalline secondary mineral phases, ultimately making the difference between the granitic and basaltic parent materials less striking.

The big mineralogical difference between basaltic and granitic soils in our study is in the amount of iron oxides, minerals which are known to coat kaolin clays and increase their specific surface area and OM stabilizing potential (e.g. Baldock and Skjemstad, 2000).

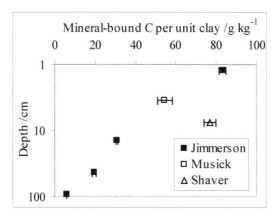

Fig. 7. Mineral-bound C normalized for clay by sampling depth. The basalt soil is represented by closed symbols and the granite soils are represented by open symbols. Standard error values are based solely on the error associated with the mineral-associated C (n=2), the error associated with % clay is unknown (n=1).

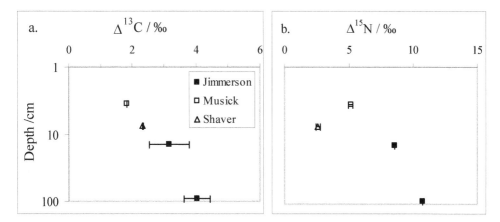

Fig. 8. Enrichment of the kaolin-dominated (2-2.6 g cm^{-3}) ^{13}C and ^{15}N values ($\Delta^{13}C=\delta^{13}C_{soil}-\delta^{13}C_{roots}$ and ^{15}N $\Delta^{15}N =\delta^{15}N_{soil}-\delta^{15}N_{roots}$) as a function of depth. The basalt soil is represented by closed symbols and the granite soils are represented by open symbols. Standard error bars reflect the simple rules for propagation of error of sums and differences.

But neither the iron-coated kaolin fractions nor the iron oxide fractions had particularly high concentrations of OM. We conclude that the effect of iron oxides in C stabilization occurs indirectly through the formation of highly stable aggregates and is also relatively more pronounced in B horizons, where iron is most abundant.

The Musick and Jimmerson sites lie just below, and the Shaver soil just above, the permanent winter snow line. The Shaver soil was the least ^{15}N-enriched at all depths,

possibly due to more effective retention of mineral forms of N or a preferential loss of organic N (Amundson et al., 2003) during snowmelt at that site. Intense leaching at snowmelt retards the accumulation of kaolinite and promotes the formation of vermiculite (Dahlgren et al., 1997), with about fifty times more specific surface area. This mineral difference led to higher levels of mineral-bound C per unit clay in the Shaver than in the Musick soil (Figure 8). This effect is not an artefact of sampling depth, which would have had the opposite effect, but may be enhanced by the direct effect of temperature on biodegradation.

Mineralogy and/or climate may also influence C turnover, which was slightly slower in the Shaver than the Musick clay fractions (142-167 versus 110-115 years, Table 7). Wattel-Koekkoek and Buurman (2004) report similarly modest differences in the cycling rates of OM on high activity smectite versus low activity kaolinite species, and we suspect that much larger differences in clay activity, such as between noncrystalline and crystalline species, are required to highlight the direct effect of mineralogy on C cycling rates (Torn et al. 1997).

4.3 The Jimmerson soil mineralogy and profile development

The mineral analysis of the Jimmerson soil revealed an absence of allophanes and the presence of quartz. The occurrence of quartz and cristobalite on this broad flat lava flow suggest that either (1) quartz grains arrived by aeolian transport from a nearby sandstone formation; or (2) the parent material, which is mapped as a broad Pleistocene vesicular olivine basalt lava flow (MacDonald and Lydon, 1972), may in fact be a basalt-andesite intergrade.

It has been observed (Allen and Hajek, 1989) that halloysite-dominated soils are rare, however the XRD data for the Jimmerson soil clearly indicate that halloysite is the main secondary mineral in both the A and B horizons. Volcanic glass and allophanes are precursors to halloysite, which can also precipitate from the desilication of smectites (2:1 layer silicates) and may subsequently recrystallize to produce kaolinite (Allen and Hajek, 1989; Hendricks and Whittig, 1968; Southard and Southard, 1989). The absence of allophanes is related to the relative age of the soil. This soil, which is not glaciated and has been undergoing weathering for thousands of years, is presumably at an advanced stage of development that lies between the allophanic (or smectitic) and kaolinitic endmembers. As noted above, this mineralogical stage has muted parent material effects on soil C storage that might be expected in less developed soils.

The depth profile of the >3 g cm^{-3} iron oxide fraction in the Jimmerson soil exhibited an intriguing pattern for which we present two interpretations: (1) The suite of organic compounds associated with the goethite and hematite may change with depth – a hypothesis that could be tested using ^{13}C NMR spectroscopy (e.g. Golchin et al.1994); and/or (2) young humus with relatively high stable isotope values may be preferentially 'cheluviated' from the A1 to the A2 of horizon. Because the crystalline iron oxides we identified in the XRD patterns are almost certainly immobile, a corollary to this hypothesis is that the > 3 g cm^{-3} fractions also captured non-crystalline iron phases, an assumption that could be tested by directly examining the ^{14}C and ^{15}N contents of OM in various

mineralogical phases separated by sodium pyrophosphate and ammonium oxalate extractions (e.g. Masiello et al., 2004).

5. Conclusions

In this study we examined three northern California forest soils: the Jimmerson (warm, basaltic), the Musick (warm, granitic), and the Shaver (cool, granitic). As expected, we found levels of clay, iron, and aggregation to be highest in the warmer basaltic soil and lowest in the cooler granitic soil. But although clay content was correlated with degree of humification as indicated by C/N ratios, there was no difference in % C depth profiles and total C storage across these three soils. In the A horizons, where most OM resides, the iron oxide fraction of the basaltic soil was associated with a very minor proportion of the OM. Overall, it was climate differences between the two granitic soils, rather than differences in the mineral composition of the parent materials, that resulted in the most obvious mineral effect on C. In the cooler granitic soil, more intense leaching has promoted the formation of more reactive clay and hence more mineral-bound C per unit clay.

The impetus of this study was to evaluate the mineral density separation method in the context of understanding the role of parent material and mineralogy in organic matter storage and turnover. Density is clearly appropriate for separating minerals with different specific gravities – such as allophanes, crystalline silicate clays, and iron oxides – and it is more effective when aggregation is weak – such as in ashy or sandy soils. In part because they span a relatively wide range of densities, primary minerals are easy to separate. They are also easy to subsequently identify. They do not, however, harbour much OM, so their utility in OM studies may be limited to special cases, such as ammonium fixation by altered mica.

Conversely, distinct silicate clay species, which are associated with high OM content, have a relatively high degree of overlap in their specific densities, especially in light of their OM as well as iron oxide coatings. For example, this study corroborates the notion that the density of organo-clay complexes clearly reflects the OM:clay ratio (as well the degree of alteration of the clay-associated OM). But, although the slowest cycling C is indeed associated with clays, as most studies indicate, so is fast cycling C, and separating these two pools requires additional chemical treatment. In conclusion, it will often be necessary to follow preliminary density separations with complementary techniques.

If certain groups of mineral species occur and function together, as is the case with kaolins and their likely ferric coating, there may be little value in attempting to separate them. In fact, our density separation may have lifted some colloidal ferric coatings from the kaolin surfaces and combined them with crystalline iron oxides, thereby obscuring our ability to distinguish the effect of these ecologically distinct species on soil OM patterns. We note, however, that the density separation technique can be used to exploit differences in particle size as well as density. Iron species can range from sand sized macro crystals to colloidal coatings and can be separated on the basis of their size – another determinant of settling velocity.

Finally, although sonication is required in order to disaggregate the constituent minerals of most soils, when the aggregation is strong it may be impractical to deliver enough energy to accomplish this. For example, the sonication of the basalt soil was insufficient to fully lift the

iron coatings off the kaolin, but excessive insofar as it resulted in a significant amount of OM becoming dissolved in the sodium polytungstate solution. This kind of trade-off needs to be evaluated on a case by case basis.

This study highlights advantages and difficulties associated with the density fractionation approach. The success of the density separation is highly dependent on the individual soil characteristics – such as aggregation and the degree of overlap in the densities of the constituent clay minerals – and on the purpose of the separation, that is how the OM attributes of interest are related to the mineral density differences. We conclude that the principle of separating intact soil into discrete organo-mineral complexes on the basis of gravity is   reproducible and effective, but the organic matter pools associate with these fractions are still heterogeneous with respect to composition and turnover time.

6. Acknowledgements

This research was undertaken as part of Cristina Castanha's doctoral dissertation and primarily funded by a Kearney Foundation of Soil Science grant to Susan Trumbore. Ronald Amundson's participation was funded by the California Agricultural Experiment Station. The ^{14}C measurements were funded by a grant to Amundson from the Center for Accelerator Mass Spectrometry at the Lawrence Livermore National Laboratory. We thank Isabelle Basile for suggestions on soil mineral and density separation, Rudy Wenk and Timothy Teague at the Department of Earth and Planetary Science at UC Berkeley for facilitating the X-ray diffraction analyses, Alex Blum at the USGS in Menlo Park, California for help in designing a low background holder for small XRD samples, Andy Thompson in the Silver Lab at UC Berkeley for help with the CN analyzer, Paul Brooks and Stefania Mambelli at the Center for Stable Isotope Biogeochemistry at UC Berkeley for their kind and indefatigable assistance with the Mass Spectrometer, Michaele Kashgarian and Paula Zermeno at the Center for Accelerator Mass Spectrometry at LLNL for assistance with ^{14}C AMS measurements.

7. References

Allardice, W.R., Munn, S.S., Begg, E.L. & Mallory, J.I. 1983. *Laboratory Data and Description for Some Typical Pedons of California Soils. Volume I: Central and Southern Sierra.* Department of Land, Air, and Water Resources, University of California Davis.

Allen, B.L. & Hajek, B.F. 1989. Mineral Occurence in Soil Environments. In: *Minerals in Soil Environments* (eds J.B. Dixon & S.B. Weed), pp. 199-264 . Soil Science Society of America, Madison, Wisconsin.

Amundson, R., Austin, A. T., Schuur, E.A.G., Yoo, K., Matzek, V., Kendall, C., Uebersax, A., Brenner, D. & Baisden, W.T. 2003. Global patterns of the isotopic composition of soil and plant nitrogen. *Global Biogeochemical Cycles*, 17:1031-1041.

Baisden, W.T., Amundson, R.G., Cook, A.C. and Brenner, D.L., 2002. Turnover and storage of C and N in five density fractions from California annual grassland surface soils. *Global Biogeochemical Cycles*, 16: 1117-1132.

Baldock, J.A. & Skjemstad, J.O., 2000. Role of the soil matrix and minerals in protecting natural organic materials against biological attack. *Organic Geochemistry*, 31: 697-710.

Barnhisel, R.I. & Bertsch, P.M. 1989. Chlorites and hydroxy-interlayered vermiculite and smectite. In: *Minerals in Soil Environments* (eds J.B. Dixon & S.B. Weed), pp.729-779. Soil Science Society of America, Madison, Wisconsin.

Begg, E.L., Allardice, W.R., Munn, S.S. & Mallory, J.I., 1985. *Laboratory Data and Description for Some Typical Pedons of California Soils. Volume III: Southern Cascade and Northern Sierra.* Department of Land, Air, and Water Resources, University of California Davis.

Boutton, T.W. 1991. Stable carbon isotope ratios of natural materials. In: *Carbon Isotope Techniques.* (eds D.C. Coleman & B. Fry), pp. 155-171. Academic Press.

Brindley, G.W. & Brown, G. (eds). 1984. *Crystal Structures of Clay Minerals and Their X-ray Diffraction Identification.* Mineralogical Society, London.

Christensen, B.T., 1992. Physical fractionation of soil and organic matter in primary particle size and density separates. *Advances in Soil Science,* 20: 1-90.

Dahlgren, R.A., Boettinger, J.L., Huntington, G.L. and Amundson, R.G. 1997. Soil development along an elevational transect in the western Sierra Nevada, California. *Geoderma,* 78: 207-236.

Ehleringer, J.R., Buchmann, N. & Flanagan, L.B. 2000. Carbon isotope ratios in belowground carbon cycle processes. *Ecological Applications,* 10: 412-422.

Feller, C. & Beare, M.H. 1997. Physical control of soil organic matter dynamics in the tropics. Geoderma, 79: 69-116.

Golchin, A., Oades, J.M., Skjemstad, J.O. & Clarke, P. 1994. Study of free and occluded particulate organic matter in soils by solid state ^{13}C CP/MAS NMR spectroscopy and scanning electron microscopy. *Australian Journal of Soil Research,* 32: 285-309.

Golchin, A., Oades, J.M., Skjemstad, J.O. & Clarke, P., 1995. Structural and dynamic properties of soil organic matter as reflected by 13C natural abundance, pyrolysis mass spectrometry and solid-state-13 C NMR spectroscopy in density fractions of an oxisol under forest and pasture. *Australian Journal of Soil Research,* 33: 59-76.

Harradine, F. & Jenny, H. 1958. Influence of parent material and climate on texture and nitrogen and carbon contents of virgin California soils I. Texture and nitrogen contents of soils. *Soil Science,* 85: 235-243.

Harradine, F.F. 1954. *Factors influencing the organic carbon and nitrogen content of California soils.* Doctoral Dissertation, University of California Berkeley.

Hendricks, C.W. & Whittig, L.D. 1968. Andesite weathering. II. Geochemical change from andesite to saprolite. *Journal of Soil Science,* 19: 147-153.

Hogberg, P. 1997. Tansley review no. 95 15N natural abundance in soil-plant systems. *New Phytologist,* 137: 179-203.

Jaynes, W.F. & Bigham, J.M. 1986. Concentration of Iron Oxides from Soil Clays by Density Gradient Centrifugation. *Soil Science Society of America Journal,* 50: 1633-1639.

Kendall, C. and Caldwell, E.A., 1998. Fundamentals of Isotope Geochemistry. In: *Isotope Tracers in Catchment Hydrology* (eds C. Kendall & J.J. McDonnel), pp. 51-84. Elsevier Science, New York.

Krull, E.S., Baldock, J.A. & Skjemstad, J.O. 2003. Importance of mechanisms and processes of the stabilisation of soil organic matter for modelling carbon turnover. *Functional Plant Biology,* 30: 207-222.

Levin, I. & Hesshaimer, V. 2000. Radiocarbon-a unique tracer of the global carbon cycle dynamics. *Radiocarbon,* 42: 69-80.

MacDonald, G.A. & Lydon, P.A. 1972. Geologic map of the Whitmore quadrangle. U.S. Geological Survey.

Masiello, C.A., Chadwick, O.A., Southon, J., Torn, M.S. & Harden, J.W. 2004. Weathering controls on mechanisms of carbon storage in grassland soils. *Global Biogechemical Cycles*, 18. doi: 10.1029/2004GB002219.

McKeague, J.A., Brydon, J.A. & Miles, N.M. 1971. Differentiation of forms of extractable iron and aluminum in soils. *Soil Science Society of America Proceedings*. 35: 33-38.

Minagawa, M., Winter, D.A. & Kaplan, I.R. 1984. Comparison of Kjedahl and combustion methods for measurement of nitrogen isotope ratios in organic matter. *Analytical Chemistry*, 56: 1859-1861.

Monnier, G., Turc, L. & Jeanson-Luusinang, C. 1962. Une methode de fractionnement densimetrique par centrifugation des mateires organiques du sol. *Annales Agronomiques*, 13: 55-63.

Nadelhoffer, K.J. & Fry, B. 1988. Controls on natural nitrogen-15 and carbon-13 abundances in forest soil organic matter. *Soil Science Society of America Journal*, 52: 1633-1640.

Shang, C. & Tiessen, H. 1998. Organic matter stabilization in two semiarid tropical soils: Size, density, and magnetic separations. *Soil Science Society of America Journal*, 62: 1247-1257.

Southard, S.B. & Southard, R.J. 1989. Mineralogy and Classification of Andic Soils in Northeastern California. *Soil Science Society of America Journal*, 53: 1784-1791.

Spycher, G. & Young, J.L. 1979. Water dispersible soil organic mineral particles. II: Inorganic amorphous and crystalline phases in density fractions of clay size particles. *Soil Science Society of America Journal*, 43: 328-332.

Stuiver, M. & Polach, H.A. 1977. Reporting of ^{14}C data. *Radiocarbon*, 19: 355-363.

Torn, M.S., Trumbore, S.E., Chadwick, O.A., Vitousek, P.M. & Hendricks, D.M. 1997. Mineral control of soil organic carbon storage and turnover. *Nature*, London, 389: 170-173.

Trumbore, S.E. 1993. Comparison of carbon dynamics in tropical and temperate soils using radiocarbon measurements. *Global Biogeochemical Cycles*, 7: 275-290.

Trumbore, S.E. & Zheng, S. 1996. Comparison of fractionation methods for soil organic matter ^{14}C analysis. *Radiocarbon*, 38: 219-229.

Wattel-Koekkoek, E.J.W. & Buurman, P. 2004. Mean residence time of kaolinite and smectite-bound organic matter in Mozambiquan soils. *Soil Science Society of America Journal*, 68: 154-161.

Mineralogy of Basaltic Material on the Minor Bodies of Our Solar System

René Duffard

Instituto de Astrofísica de Andalucia – CSIC, Granada, Spain

1. Introduction

The main idea of this chapter is to study the presence of basaltic material in the minor bodies of our Solar System. Basaltic material in the solar system is found not only in the surface of the terrestrial planets, but also on the surface of the Moon and asteroids. This material is reckoned as the result of an extensive geochemical differentiation.

Our modern understanding of the formation of the solar system is that it began in a cold cloud of gas and dust under conditions similar to other nebulae. After the dissipation of the gas and dust, the nebula continues to evolve as the planetesimals form larger bodies. In the process, these larger bodies are heated by collisions and radioactive decay. Asteroids are the remnants of these planetesimals that could not form a bigger planet. The materials currently present in the asteroids are the most primitive, less evolved one, and that's why is important to understand.

From this material, the less evolved is called chondritic and is composed with the same material (without the volatiles) as our star, the Sun. After accretion, the parent body consists of primitive material and is heated internally by decay energy of short-lived nuclides such as ^{26}Al (with a half-life of 0.72 Myr) and ^{60}Fe (with a half-life of 2.6 Myr). Some planetesimals accreted to form bigger objects but no thermal evolution was present and today the original chondritic material can be observed. Some other planetesimals, have a most complicated thermal evolution. Partial fusion of the initial chondritic material makes, first, the heavier liquid (Fe-Ni-S) migrate to the centre and second, the lighter silicated liquid (SiO_2) migrates to the surface. The final result of such a process is a body with a dense metallic core, a mantle of lighter olivine-rich material and an even lighter basaltic surface. This structure is seen in the Earth, the Moon, and terrestrial planets (Figure 1).

There is one particular case in which the original planetesimal formation of a differentiated small body can be observed. Initially, remote observations using ground-based telescope showed that the surface of the asteroid (4) Vesta is composed of basaltic material. Vesta with a mean diameter of 530 Km and a basaltic surface was an excellent candidate to be a small terrestrial planet. After the discovery of the basaltic crust on Vesta, the idea of a proto-planet, that could not grow to a bigger planet but was differentiated start to be accepted in the community.

Fig. 1. Chronological step of the formation of a differentiated planetesimal. From left to right, the agglutination, heating, differentiation in a Fe-Ni nucleus, silicate mantle and basaltic crust and finally the identification with different kinds of meteorites.

Then, more asteroids' surfaces showed the same composition. These smaller asteroids (kilometre sized) were detected in the same region of Vesta and later were proved that they come from a craterization event on Vesta's surface. Then, asteroids with near Earth orbits (NEAs) were also identified to have basaltic surfaces. Later, the relation between a rare classes of meteorites (Howardites-Eucrites-Diogenites, HED) and these asteroids were linked together and the problem seems to be solved (meteorites are pieces of asteroids found on the surface of the Earth).

The proposed scenario was that great impacts excavated the surface of Vesta, producing a swarm of small fragments. Part of them, were injected into dynamical resonances which pumped up their orbital eccentricities, and were thus ejected due to close encounters with terrestrial planets. Most of these fragments fell directly into the Sun or escaped from the Solar System, but part of them remained in near-Earth orbits. Further collisions ejected fragments into Earth-colliding orbits, becoming the HED meteorites recovered on Earth (Drake 2001). Several observational facts corroborate this scenario: the identification of a Vesta dynamical family (Williams, 1989; Zappalá et al., 1990) called Vestoids, the confirmation that small asteroids in the region near (4) Vesta do have a basaltic surface composition (Binzel and Xu, 1993; Burbine et al., 2001) called V-type asteroids, the identification of a large impact basin on (4) Vesta (Thomas et al., 1997), the discovery of several NEAs with basaltic mineralogical surface composition (McFadden et al., 1985; Cruikshank et al., 1991; Binzel et al., 2004; Duffard et al., 2006) and, last but not least, the identification that most of the HED have similar isotopic composition, indicative of a common origin (Clayton and Mayeda, 1983, 1996; Mittlefehldt et al., 1998). The problem seems to be solved, and all the basaltic material was coming from the surface of Vesta, the only large enough asteroid to be differentiated.

But, there are few very important observations, however, that do not readily fit into an all-encompassing Vesta-HED story. A small (~30km diameter) basaltic asteroid, (1459) Magnya, was identified in the outer asteroid belt (Lazzaro et al. 2000). Magnya was proved not to come from Vesta. It was the first one, and then followed by the discovery of several others basaltic asteroids, all of them not related to Vesta (Duffard and Roig, 2009; Moskovitz et al. 2008).

On the other hand, in the field of meteorites, more precise laboratory techniques allows to identify that some HEDs are different from the main group. Meanwhile, more basaltic meteorite findings showed that some of them are completely different from the HEDs in the oxygen isotopic plot (Yamaguchi et al. 2002, Bland 2009.).

In Figure 2 is plotted all the basaltic material identified in the minor bodies, principally on the Main Asteroid Belt. On this plot of semi-major axis versus eccentricity of the orbits, Vesta is in the middle of the family cloud (black dots). The Vesta family is identified as fragments that can be dynamically linked with the big asteroid. Then Vestoids (blue squares) are kilometre sized asteroid, identified to have a basaltic surface and out of the family limits. The identification is done taken a reflectance spectrum of the surface (Florckzak et al. 2002). The origin of these objects should be Vesta but they suffer a dynamical process, which change the orbit and took it out from the limits of the family (Carruba et al. 2005). Magnya (at 3.18 AU) and several other small basaltic asteroids in the outer main belt (semi-major axis bigger than 2.5 UA) are all not related to Vesta. Finally, the basaltic SLOAN candidates are marked as green squares. The Sloan Digital Sky Survey Moving Objects Catalogue (SDSS-MOC; Ivezic et al., 2001; Juric et al., 2002) was used to identify candidate V-type asteroids. The SLOAN photometric survey obtained the photometry in 5 different wavelengths of several thousands asteroids. From that, the identification of the V-type candidates is done and at the end, confirmed spectroscopically (Duffard and Roig, 2009).

In summary, the asteroid Vesta remains the only big intact basaltic asteroid. Vesta has a swarm of near 4000 objects that are fragments from collisions on its surface. These small asteroids are fragments of the crust of Vesta and can give us information on different depths of the crust. Studying these vestoids give us clues on the different depths and is the only way to study the interior of a body of that size.

Although, both the smaller Vestoids and Magnya are clearly basaltic in nature, the Vestoids are related to Vesta, but Magnya and other in the outer main belt are not. Meteorites HEDs, NWA011, and Ibitira have very similar basaltic mineralogy, but appear to have distinct parent bodies, as it will be shown in next section. As shown in Figure 2, basaltic material is not only present in the Vesta region, but also in different parts of the Main Asteroid Belt. All the small asteroids found that are not related to Vesta are small (less than 20 km) and should be pieces of the crust of a larger differentiated object. Still not answered questions are: if the pieces of the crust are observed today, why the pieces of the mantle or nucleus of that body are not observed? If the body that generated that fragments was not completely destroyed, why it is not found? More searches looking for basaltic material not related to Vesta is in a urgent need.

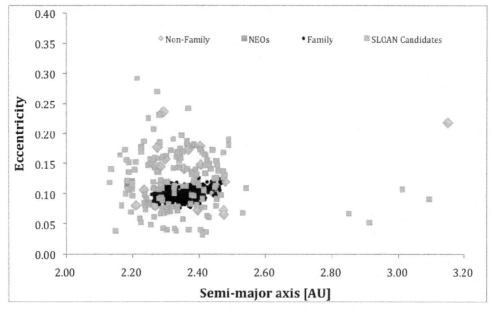

Fig. 2. Semi-major axis vs. eccentricity for the entire sample of basaltic asteroids identified until now.

2. Basaltic meteorites

One criterion to classify the meteorites is in two big groups: **chondritic** and **achondritic**. The chondritic meteorites are samples of the most primitive material, the less altered (thermal, aqueous, radiated, etc). They present chondrules that are millimetre-sized spheres formed in the initial stage of the Solar System.

Achondritie meteorites are those that suffered extreme metamorphosis from the original chondritic material. Achondrite meteorites are divided in **Primitive Achondrites** (Acapulcoites, Lodranites, Brachinites, Winonaites, Urellites and the ungrouped Primitive Achondrites), **the HED group** (Howardites, Eucrites, and Diogenites), the other **evolved asteroidal achondrites** (Angrites and Aubrites), **the Lunar group** and finally the Martian meteorites called **SNC group** (Shergottites, Nakhlites and Chassignites). All these objects are Stony meteorites. On the other hand, there are the Stony-iron and Iron meteorites that represent the transition between mantle-nucleus and nucleus, respectively.

The Eucrites are named for a Greek word meaning "easily distinguished". Representing the most common class of achondrites, more than 100 Eucrites are known, excluding all probable pairings. Although they are easily distinguished from chondrites, they closely resemble terrestrial basalts. Actually, Eucrites are extraterrestrial basalts; volcanic rocks of magmatic origin, representing the crust of their parent body, thought to be Vesta. They are primarily composed of calcium-poor pyroxene, pigeonite, and calcium-rich plagioclase, anorthite. Additionally, Eucrites often contain accessory minerals such as silica, chromite, troilite, and nickel-iron metal. Based on mineralogical and chemical differences, the Eucrites have been further divided into three distinct subgroups: the non-cumulate group, the

cumulate group, and the polymict group. In the microscope, they resemble terrestrial lava flows, and for a long time cosmochemists figured that Eucrites formed in the same way terrestrial basalts form, by partial melting of the interior of the parent body. However, assorted geochemical arguments, especially the concentration of siderophile elements (they concentrate in metallic iron), suggest that the parent body was totally melted (or nearly so) when it formed. As it crystallized, the last 10-20% of magma would have a composition like the Eucrites. Of course, somehow that magma has to be squirted onto the surface to make lava flows. Nevertheless, it is clear that Eucrites formed as lava flows. Their ages indicate that it happened 4.5 billion years ago.

The Diogenites are rare and this group consists only of about 40 members if all probable pairings are excluded, especially those that have been found in the ice fields of Antarctica. They are named for a Greek philosopher of the fifth century B.C., Diogenes of Apollonia. Mineralogically, the Diogenites are composed primarily of magnesium-rich orthopyroxene, with only minor amounts of olivine and plagioclase. The pyroxenes are usually coarse-grained, suggesting a cumulate origin for the Diogenites in magma chambers within the deeper regions of parent body's crust. They are intrusive igneous rocks similar to plutonic rocks found on Earth, and they experienced much lower cooling rates than did the Eucrites, which allowed the pyroxene to form sizeable crystals.

Howardites are named for Edward Howard, a renowned British chemist of the 18th century and one of the pioneers of meteoritics. They are nearly as rare as Diogenites, and there are only about 50 members to this group if all probable pairings are excluded. Consisting primarily of eucritic and diogenitic clasts and fragments, Howardites are polymict breccias. However, they also contain dark clasts of carbonaceous chondritic matter and impact melt clasts, indicating a regolith origin for the members of this group. The Howardites represent the surface of the parent body, a regolith breccia, consisting of eucritic and diogenitic debris that was excavated by a large impact. A schematic model of four hypothetical impacts into the layered crust producing different types of Eucrites is shown in Figure 3.

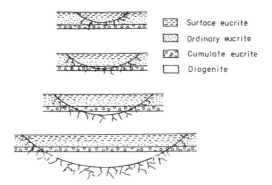

Fig. 3. A model of the layered crust of the HED parent body. (Figure from Takeda, 1997).

An important tool to identify the parent body of a group of meteorites like the HEDs is the oxygen isotopic plot. Oxygen has three stable isotopes with mean terrestrial abundance

ratios: $^{17}O/^{16}O = 1/2700$ and $^{18}O/^{16}O = 1/490$. These ratios are variable in natural materials due to a variety of physical and chemical processes, ranging from stellar nucleosynthesis to ordinary mass-dependent chemical fractionation. As a natural isotopic tracer, oxygen has an advantage over other light elements, such as hydrogen, carbon, and nitrogen, in that the combined variation of two isotope ratios ($\Delta^{17}O = {}^{17}O/^{16}O$ and $\Delta^{18}O = {}^{18}O/^{16}O$) helps to identify the underlying process (Clayton 2002).

Chemical and physical processes within an isolated planetary body, such as the Earth, Moon, and Mars, almost always obey a simple mass-dependent relationship between variations in $\Delta^{17}O$ and $\Delta^{18}O$ ratios which is nearly 0.52. All rocks from the Earth should plot in a line of slope 0.52 and this line is called the Terrestrial Fractional Line (TFL). All meteorites from the same parent body should plot in a line with the same slope but parallel to the TFL. This oxygen isotopic plot helps us to identify if a meteorite can came from the same parent body as other one.

This is what it can be seen in Figure 4 where the slope of 0.52 is taken out and several HED meteorites plotted out of the horizontal line. Most of the Eucrites and Diogenites plot on the same line (the Eucrite Fractional Line – EFL) but there are several other Eucrites that plots outside the line, indicating the different origin.

From the basaltic meteorites, there are at least pieces from seven different crusts. On the basis of its unusual oxygen isotope composition, NWA 011 (out of the plot) is presumed to derive from a basaltic parent asteroid other than (4) Vesta. Six anomalous Eucrite-like basaltic meteorites were identified on the basis of their oxygen isotope compositions (and, in some cases, unusual mineralogy): Ibitira (a meteorite that fell in Minas Gerais, Brazil, in 1957), Asuka-881394 (found in the Queen Maud Land region of Antarctica, 1988), PCA 91007 (found in the Pescora Escarpment region of Antarctica, 1991), Pasamonte (fell in New Mexico, USA, 1933), NWA 1240 (found in the Sahara, 2001) and Bunburra Rockhole, some of them closer to the Angrite Fractional Line - AFL.

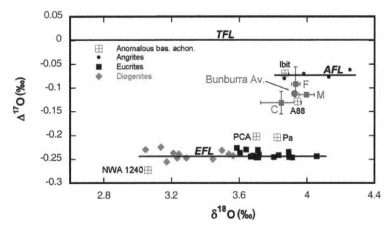

Fig. 4. Oxygen Isotopic plot showing the TFL and most of the Eucrites. Figure taken from Bland et al. (2009).

As was mentioned in the previous section, if a body suffered a differentiation, its internal structure is divided in a Fe-Ni nucleus, then a silicate (mostly olivine) mantle and a fine crust. If this kind of body is completely destroyed you can have samples of the nucleus (iron meteorites), the mantle (stony-iron meteorites) and the crust (basaltic meteorites).

From the iron meteorite collection it's proved the existence of at least 60 different nucleus (Bottke et al. 2006). This mean that at least 60 different differentiated asteroids were completely destroyed and some material were found in the surface of the Earth.

In summary, differentiation seems to be normal in the early solar system. The principal heater to produce the differentiation is the decay of ^{26}Al that has a half-life of 0.7 My. In that case, all the differentiated planetesimals formed in the first 5 My of formation of the Solar System. There are several works modelling Vesta type formation reaching that conclusion (Gupta and Sahijpal 2010 and references therein). After the formation of this small (400 Km or less in diameter) differentiated planetesimals, bigger differentiated planets start to be formed and Mars (after 12 My) or the Earth (after 32) were formed (Kleine et al. 2002). From the meteorite suite (iron meteorites) it can be inferred that at least several dozens of differentiated asteroids were formed.

3. Vesta, vestoids and V-type asteroids: Mineralogy

In the previous sections the basaltic material coming from Vesta was introduced. Until 10 years ago, it was a general thought that all the basaltic meteorites and all the basaltic asteroids are coming from this asteroid. Starting with the discovery that the surface of Magnya is composed of basalt, and then discovering several others basaltic asteroids not related to Vesta, those facts showed that Vesta is not unique. More precise laboratory techniques and meteorite founds showed the same evidence, there were more differentiated parent bodies in the beginning of the Solar System.

In this section it will be shown different techniques to analyze the spectroscopic features in the reflectance spectra of basaltic asteroids and compare with the meteorites. To obtain information on the mineralogy of asteroids, remote sensing using telescopes is a powerful technique. The most common and expanded way to obtain mineralogical information on the surface of an asteroid is the reflectance spectroscopy in the visible and near-infrared (VNIR). In this spectral range (0.4 – 2.5 microns) there are several absorption bands that can be identified to characterize the mineral/s present in the surface of the asteroid. With a suitable spectrograph in a medium size telescope, the reflectance spectra of the asteroid can be obtained. As the asteroid only reflects the light from the Sun, the final spectra must be corrected extracting the Sun light component. This final result has only the information on the asteroid's surface.

One interesting technique is to compare meteorites spectra with asteroid's spectra. As the asteroids are only accessible by remote sensing, like the reflectance spectra, lets do the same with the meteorites. The Reflectance Laboratory (RELAB) is one of the large existing databases of reflectance spectra (Pieters 2004). One of the best characteristics is the homogeneity in the spectra acquisition. In RELAB database it can be found reflectance spectra of meteorites and minerals from 0.3 to 2.6 microns, in same cases taken with different grains sizes, temperatures, and other conditions.

The first step to identify the mineralogy of an asteroid is the comparison of the reflectance spectra of the asteroid with one taken from the reflectance spectra databases (Figure 5). All the basaltic reflectance spectra are characterized by the presence of prominent absorption bands, near 1 and 2 µm, indicative of a mixture of pyroxene and possibly olivine. The spectrum of pyroxene is dominated by its characteristic absorption features near 1.0 and 2.0 µm. Olivine spectrum is dominated by a complex absorption centred near 1.0 µm. In both cases, these absorptions are produced by spin-allowed electronic transition of Fe^{2+} in distorted octahedral (M1 and M2) crystal field sites (Burns, 1970).

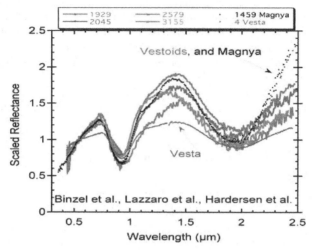

Fig. 5. Reflectance spectra of asteroid (4) Vesta, Magnya and some Vestoids. It can be seen the two characteristics absorption bands near 1 and 2 microns. Figure taken from Pieters et al. 2005.

3.1 Mineralogical analysis

One of the most important parameters in order to characterize the mineralogy associated with this kind of spectra is the position of the centre of the absorption bands near 1 and 2 µm. Two parameters can be used: the *band minimum* and the *band centre*. These are defined (Cloutis and Gaffey, 1991) as the wavelength position of the point of lowest reflectance before and after the removal of the continuum, respectively.

The spectra obtained after the removal of the continuum were used to compute the position of the *band centres*, using the same procedure adopted for computing the minima and maxima. These final spectra were used also to compute the *band areas* and the *band depths*. Following Cloutis (1985), *band area* is defined as the area enclosed by the spectral curve and a straight-line tangent to the respective maxima. By convention *band area I* is between the maxima at 0.7 and 1.4 µm, while *band area II* is between the maxima at 1.4 and 2.4 µm. The parameter *BAR* is then defined as the ratio between the *band area II* and the *band area I*. Finally, the *band depth I* and the *band depth II* are measured as the ratio of the reflectance maximum at the peak between *band I* and *band II* to the reflectance of the *band I centre* and to the reflectance of the *band II centre*, respectively.

Each of the above computed spectral parameters are diagnostic of the associated mineralogy present on the surface of the observed asteroids. The relationship between these parameters and the mineralogy for reflectance spectra, particularly pyroxene and olivine, has been studied in various papers over the last years (Adams, 1974, 1975; King and Ridley, 1987; Cloutis and Gaffey, 1991) and recently reviewed by Gaffey et al. (2002) and Duffard et al (2004, 2005).

As stated above, *band centres* are among the most important diagnostic parameters of the mineralogy in a spectrum. According to several authors (Adams, 1974; Cloutis and Gaffey, 1991) in most pyroxenes and in the basaltic achondrites there is a strong correlation between the position of *band I centre* and *band II centre* and the associated mineralogy. For example, orthopyroxene bands shift to longer wavelengths with increasing amounts of iron, whereas clinopyroxene bands shift to longer wavelength with increasing calcium content. The parameter BAR, which is the ratio between the *band areas* II and I, can give information about the abundance of olivine since for pure olivine this parameter is essentially zero for all the phases.

In figure 6 its shown the relationship between the BAR parameter and band I centre for a sample of achondrite meteorites. In this plot its marked the regions where the meteorites with high content of olivine (Ol), the ordinary chondrites (OC) and the basaltic asteroids (BA) should plot (right panel). In this plot, HED meteorites and some other achondrites are plotted. In the right panel it's shown the same areas, but V-type asteroids and some other minerals are plotted. Due to the large error in the BAR parameter determination for Magnya, this asteroid plots in an extended region.

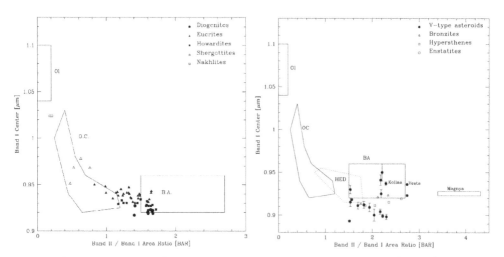

Fig. 6. BAR parameter versus band I centre for the achondrite meteorites: Howardites, Eucrites and Diogenites (left panel) and for some V-type asteroids. Figure from Duffard et al. 2005.

The pyroxene composition, i.e., the molar calcium, [Wo], and iron, [Fs], contents can be obtained from the values of the absorption band centres near 1 and 2 μm using equations

derived by exhaustive laboratory calibrations performed by several authors (Adams, 1974, 1975; King and Ridley, 1987; Cloutis and Gaffey, 1991). These equations have recently been recompiled by Gaffey et al. (2002) and in order to derive the pyroxene composition they are used in an iterative mode.

This relation was, therefore, used to compute the relative abundance of the [Wo] and [Fs] contents, as shown in Figure 7. In this figure it can be seen the determination of [Wo] and [Fe] content for all the asteroids determined in this work and from literature (Duffard et al. 2004, deSanctis, et al. 2011). The measured ranges of the average [Fs] and [Wo] contents of pyroxene (Mittlefehldt et al. 1998, Takeda 1997) are for the Eucrites (in red): [Fs] ~ 30–55 and [Wo] ~ 6–15; Howardites (in blue): [Fs]~31–42 and [Wo]~4–8; Diogenites (in green): [Fs]~20–30 and [Wo] ~ 1–3. Asteroidal values using the mentioned equations are in the range of HED meteorites, considering the large errors (+/- 5) in the determination of [Fe] and [Wo].

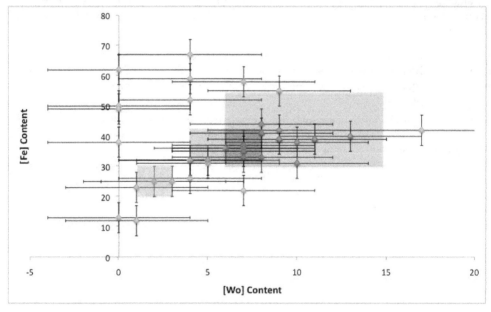

Fig. 7. Molar contents for the observed V-type asteroid obtained from Duffard et al. 2004 and DeSanctis et al. 2011.

In conclusion, a large and homogeneously obtained set of data redefines the regions where the different meteorite classes plot in spectral parameters spaces. Moreover, it is possible to study how these regions can be displaced by different physical properties, such as grain size, temperature, and mineralogy. However, when applying this kind of plot to the specific problem of relating classes of meteorites to types of asteroids, it can be observed that larger variations than expected do appear. This indicates either that it is indeed plotted different mineralogies or that this kind of plot is not appropriate for such a study. The second hypothesis is more favourable (Duffard et al. 2005). This kind of plot is only indicative on the mineralogy present on the surface of the asteroid. Another method is necessary to confirm or refute the minerals present on the asteroid. Of course, more data on meteorites

and asteroids is needed before a secure conclusion is reached and the genetic link between specific members of these two populations is definitively established.

3.2 Modified Gaussian Model (MGM)

Another powerful technique to determine the mineralogy of a remote observed asteroid is the analysis using the Modified Gaussian Model (Sunshine et al. 1990). The Modified Gaussian Model (MGM) arose from the necessity of a more general fitting method of analysis that could resolve and distinguish individual spectral absorption features and representing them with discrete mathematical distributions. The use of this quantitative correlation is only dependent of the spectrum itself. MGM method supplies an objective and consistent tool to examine the individual absorption features of a spectrum (Sunshine at al. 1990).

This method uses deconvolution to represent absorption bands as discrete mathematical distributions (modified Gaussians) taking into account the physical processes that lead to the formation of the bands. Each individual absorption band is defined by three parameters: centre, width and strength. From previous analysis (Section 3.1) it's known that the surface of basaltic asteroids could be mainly composed of pyroxene (as they are classified as V-type asteroids). Following Sunshine and Pieters (1993), it can be computed the proportion of orthopyroxene or Low Calcium Pyroxene (OPX or LCP) and clinopyroxene or High Calcium Pyroxene (CPX or HCP) present in the surface of the observed asteroids. Eight individual absorption bands were used for each spectrum, corresponding to the individual absorption bands of both end-members. Sunshine and Pieters (1993) showed that the band centres of the primary absorptions remain essentially fixed; at 0.91 and 1.83 µm for OPX, and 1.02 and 2.29 µm for CPX, and that only the relative band strengths change as a function of CPX and OPX abundance. This effect can be quantified by computing the "Component Band Strength Ratio" (CBSR) defined as the ratio between the band strengths of OPX and CPX components at 1 and 2 µm. This CBSR should be the same in the 1 and 2 µm region if the fit is correctly done with both pyroxenes.

Given a set of different MGM input parameters (individual absorption band centres, widths and strength) and operating the first step of the fitting process, it must be met strict sequential steps in order to reach physically coherent results. The calibration procedures, applied to the whole sample can be summarized as follows. Given the input parameters the first step is to proceed to an initial fit. Once the fit-result is obtained then is necessary to proceed to check the bandwidth calibration and assure it is within the tabulated values (Sunshine and Pieters 1993). All the values of bandwidth to respect of band centres found may fall within a similar and expected region. Band centre calibration follows and again, comparing the results with the tabulated values, it should confirm if the LCP/HCP ratio band near 1 and 2 µm (the component band strength ratio,the CBSR) region is acceptable. All band centres obtained should fall within the expected area of the calibration data (Adams 1974). If all these conditions are met then the final result is achieved. The individual absorption bands all combined can describe the complex absorption bands, if not then a change in the band parameters (centres, widths, strength) is needed and the process should start again. An important addition to this procedure is to always keep in mind the fit of the residual and maintain its structures to a minimum (Canas et al. 2008).

In figure 8 (left) it's shown the spectra of Eucrite meteorite Juvinas, taken from RELAB, the individual absorption bands for LCP and HCP, the combined fitted spectra superimposed to the meteorite spectra, and the extracted continuum. In the upper part of the plot it's shown the residual that is the difference between the spectra and the fitted one. No structure or band is seen in the residual. Individual's bands of the HCP and LCP are marked in the plot. In the right panel, a similar fit is shown for the basaltic asteroid 2003 YG118. The visible part of the spectra (0.5 – 0.9 μm) is taken with one instrument and the near infrared part (0.8 – 2.5 μm) with another instrument. Both spectra are joined using the mutual region near 0.9 μm.

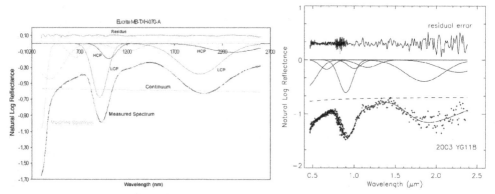

Fig. 8. An example of a MGM fit for the Eucrite Juvinas (MB-TXH-070-A) taken from the RELAB database) on the left panel and for the basaltic asteroid 2003 YG118.

After the acquisition of the spectra in the telescope, when the spectra are calibrated and ready to be analyzed, the MGM technique is applied. The first step is to verify the spectral analysis applying it to a sample of Eucrites/Diogenite meteorites. Then, the technique should be extended to a set of V-type asteroids.

Sunshine and Pieters (1993) derived a relationship from spectra of a set of powders of known proportions of high- and low-calcium pyroxenes and this relationship can be used to separate meteorite classes that have undergone various degrees of igneous processing (Sunshine et al. 2004).

3.2.1 MGM on meteorites

To build confidence in our results and to provide a basis for comparison to asteroid spectra, the first step is to examine the spectra of 25 Eucrites and 10 Diogenites (obtained from RELAB database) using MGM. These kinds of reference spectra were selected due to the spectral similarity and the higher signal to noise ratio in the meteorite spectra. On the other hand, the application of the MGM analysis in V-type spectra was recently used on the Eucrite Bouvante and V-type asteroid (4188) Kitezh (Sunshine et al. 2004), in asteroid (4) Vesta (Vernazza et al. 2005), in five Eucrites (Mayne et al. 2006), 2 V-type NEOs (deLeón et al., 2006) and 3 another V-type NEOs (Canas et al. 2008).

The MGM fit to the Eucrite meteorites indicates the presence of both low- and high-calcium pyroxene but does not require olivine to fit the overall spectrum. Similarly, the MGM fit to

the Diogenite meteorites indicates the presence of only LCP pyroxene. The fitting procedure was done as described before and in all cases the restrictions were reached. In figure 9 it is shown the band centres and band widths in function of wavelength for the sample of selected Eucrite (open circles) and Diogenites (black circles) meteorites. As can be see in the figure 9, for all the fittings the individual's bands centre of the LCP and HCP end-members

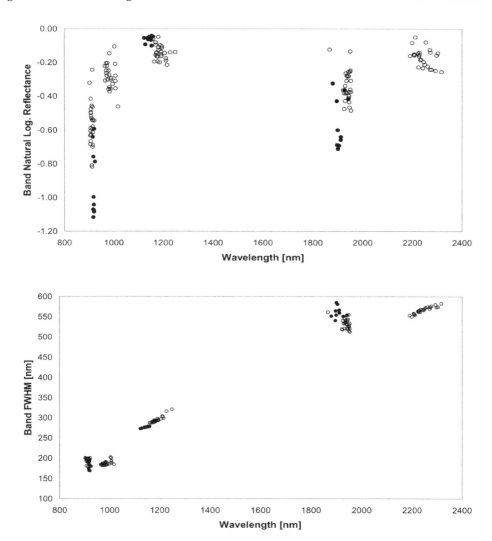

Fig. 9. Band strength for the five bands fitted to the Eucrite meteorites (white circles) and at 0.9, 1.2 and 1.9 microns for the Diogenite (black circles) meteorites (upper panel). Band widths for the same wavelengths for the Eucrites (white circles) and three bands for the Diogenite (black circles) meteorites.

are almost in the same wavelength. The individual band widths are constant for all cases. The fulfilment of this two constrains is the first indication that the fit is correct, and then the similarity of the CBSR1 and CBSR2 is the next constraint to be fulfilled. As can be seen in Table 1 the CBSR1/CBSR2 are all similar to the unity, as it was demanded in the fit. In the case of Diogenite meteorites spectra, where there is no possible CBSR determination, only the individual band centres and widths were taken into account as indicators of a correct fit.

In the case of the Eucrites, with the CBSR value it is possible to obtain the HCP/(HCP+LCP) ratio using the calibrations presented as explained in the next section. Introducing these values in the melting model presented in Sunshine et al. (2004) it is possible to obtain the melting percentage. All the values are shown in Table 1 for all the Eucrite meteorite of the sample. The final values of HCP/(HCP+LCP) and the corresponding melting percentage can be compared with the literature (Sunshine et al. 2004, Vernazza et al. 2005, Mayne et al. 2006, deLeón et al., 2006, Canas et al. 2008). Due to the excellent quality and high signal to noise ratio of the spectra, the fitting procedure was straightforward. The next expected step is to do a similar process with the asteroid spectra.

Meteorite Name	RELAB file	CBSR1	CBSR2	CBSR1/CBSR2	HCP/(HCP+LCP)	% Melting
Ibitira	MP-TXH-054-A	2.315	2.174	1.065	0.41	26
Milbillillie	MB-TXH-069-A	2.185	2.130	1.026	0.43	25
Juvinas	MB-TXH-070-A	2.619	2.576	1.017	0.37	28
Y-74450	MB-TXH-071-A	5.206	5.078	1.025	0.21	39
Padvarninkai	MB-TXH-096-C	1.539	1.478	1.041	0.57	21
Stannern	MB-TXH-097-A	1.705	1.778	0.959	0.51	22
GRO 95533	MP-TXH-066-A	1.755	1.647	1.065	0.52	22
EET A79005	MP-TXH-072-A	2.994	3.073	0.974	0.31	31
EET 87542	MP-TXH-075-A	1.674	1.681	0.996	0.52	22
EET 90020	MP-TXH-076-A	1.917	1.904	1.007	0.47	23
LEW 85303	MP-TXH-078-A	1.647	1.647	1.000	0.53	22
PCA 82502	MP-TXH-080-A	1.694	1.656	1.023	0.52	22
Cachari	MP-TXH-084-A	1.909	1.898	1.006	0.48	23
Moore County	MP-TXH-086-A	2.471	2.551	0.969	0.37	28
Pasamonte	MP-TXH-087-A	2.507	2.456	1.021	0.38	27
Bereba	MP-TXH-089-A	1.895	1.863	1.017	0.48	23
Bouvante	MP-TXH-090-A	1.701	1.644	1.035	0.53	22
Jonzac	MP-TXH-091-A	2.242	2.204	1.017	0.40	26
Serra de Mage	MP-TXH-092-A	2.443	2.522	0.969	0.37	28
A-881819	MP-TXH-096-A	2.441	2.442	1.000	0.38	27
Y-792510	MT-TXH-041-A	1.698	1.682	1.010	0.52	22
Y-792769	MT-TXH-042-A	1.731	1.765	0.981	0.51	22
Y-793591	MT-TXH-043-A	1.871	1.839	1.018	0.49	23
Y-82082	MT-TXH-044-A	1.413	1.405	1.005	0.58	20
NWA 011	MT-TXH-059	1.447	1.455	0.994	0.58	20

Table 1. Results using the MGM and the Eucrite meteorites.

3.2.2 V-type asteroids

Considering the success in modelling laboratory spectra of differentiated meteorites, the same technique is applied to asteroid spectra. As was mentioned, the reflectance spectra of V-type asteroids considered in this work were taken from different works. Spectra were obtained with different telescopes and instruments and have lower signal to noise ratio than the reflectance spectra of meteorites. The procedure of the MGM fitting in the Eucrites, was used as a guide to the fitting procedure in asteroids.

The procedure started the fitting with the same input parameters as in the Eucrite corresponding to 25% of HCP and 75% of LCP. As in the meteorite cases the main idea is to fulfil the restrictions of band centres and band width at the same time to look for the CBSR1/CBSR2 ratio equal unity. In some MGM fitting the constraints were reached without problems and it was possible to obtain the HCP/(HCP+LCP) ratio and then the melting percentage.

In the specific cases of (809) Lundia, (2468) Repin, (2763) Jeans, (2851) Harbin, (3155) Lee, (4796) Lewis, (6331) 1992 FZ1, (10037) 1984 BQ, (10285) Renemichelsen, and (10349) 1992 LN it was not possibly to obtain a reliable fit using a combination of LCP and HCP individuals bands. The fulfilment of the constraints was impossible in those cases. It was found that it cannot be reached the mentioned restrictions specifically the CBSR1 equals CBSR2. In those cases it was tried another fitting only using the LCP bands reaching a better result. It has to be mentioned that in the cases of 6331, 10285 and 10349 it was used only the infrared part of the spectra from 0.8 to 2.5 µm.

From the point of view to know something on the possible origin and related mineralogy, the sample was separated in family and non-family members and NEOs. Family members have the certainty that were ejected from the Vesta's crust. Non-family members that are in the region of similar orbital parameters that Vesta should be ejecta from the crust but can be fragments of the crust of another differentiated parent body in the region, like Eunomia o Baptistina. Near Earth Objects, do not have a precise determined origin in the Main Belt. NEO's reach they actual position after a chaotic way to the interior solar system. Most probable came from Vesta but the probability to come from another part of the Main Belt is not zero.

In Figure 10 it's shown the obtained band strength and band FWHM versus individual band centres for the asteroids. Compared with the same plot for the Eucrites (Fig. 9) the dispersion in the parameters is higher but still maintain the tendency of having similar band centres and constant band width (similar to meteorites). For the asteroids that can be obtained the HCP/(HCP+LCP) ratio it was possible to calculate the melting percentage as was made with Eucrites. The results are shown in Table 2.

The ratios of LCP/HCP bands (which are a measurement of the ratio of the band strengths) from the fitted samples were used to perform a logarithmic transformation that allows us to determine HCP/(HCP + LCP) as can be seen in Figure 11.

Based on the pre-elaborated systematic variation in the relative strength of pyroxene absorption function of the HCP/(HCP + LCP) ratios the results are related to melting percentage. Using calculations of melting with the MELTS program (Ghiorso and Sack 1995;

Asimov and Ghiorso 1998) it was determined the correspondent values of degree of melting for the samples. H chondrite material is used as starting point because is the less altered and comparable with the material present in the solar nebula. In Figure 12 is presented the plot of the HCP/(HCP+LCP) ratios in the solid residual and crystallized partial melt from an H chondrite precursor as function of percent melting (adapted from Sunshine et al. 2004). The Eucrites shaded rectangular box is from the model. Individuals regions are shown as a Eucrite polygonic region on the curve for the entire determined Eucrite meteorite sample from this work and the V-type asteroids, distinguishing that belonging or not to the dynamical family and NEOs.

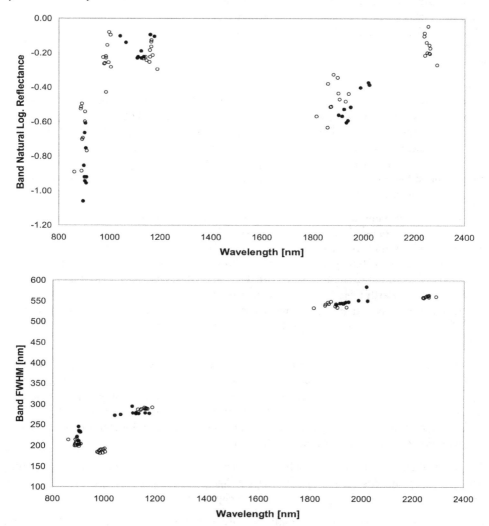

Fig. 10. Band strength (upper panel) and Band width (lower panel) for the five bands fitted to the asteroidal sample.

Asteroid Number	Family	CBSR1	CBSR2	CBSR1/CBSR2	HCP/(HCP+LCP)	% Melting
809	NO					
956	NO	3.38	3.11	1.088	0.30	32
2045	YES	3.16	3.13	1.009	0.30	32
2468	YES					
2763	NO					
2851	NO					
3155	YES					
3268	YES	2.28	2.23	1.023	0.41	26
3498	YES	3.20	3.19	1.004	0.31	31
4434	NO	2.49	2.40	1.036	0.38	27
4796	NO					
4815	YES	5.68	5.70	0.996	0.19	40
6159	YES	2.25	2.12	1.061	0.42	25
6331	YES					
6611	NEO	2.45	2.62	0.936	0.38	16
10037	YES					
10285	YES					
10349	YES					
88188	NEO	2.62	2.72	0.964	0.36	15
2003 YG118	NEO	2.02	1.89	1.069	0.46	17

Table 2. Results for the asteroidal sample.

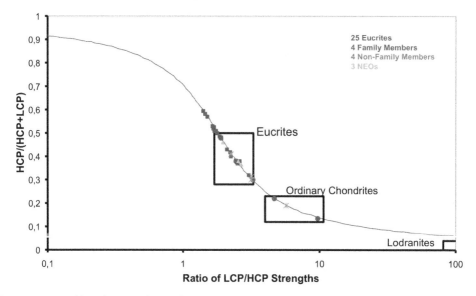

Fig. 11. Ratio of band strengths to obtain the relative proportion of LCP to HCP. Figure adapted from Sunshine et al. (2004)

To summarize, this section can be finished with some conclusions: the MGM fitting was applied to a set of 25 Eucrite and 10 Diogenites from the RELAB and this database was used for calibrating the method to be applied in the set of remotely obtained reflectance spectra of 20 V-type asteroids. This is the first time that a high number of MGM fit for Eucrite, Diogenite and V-type asteroids was calculated. In previous works, a direct comparison between spectra from meteorite and asteroids were applied.

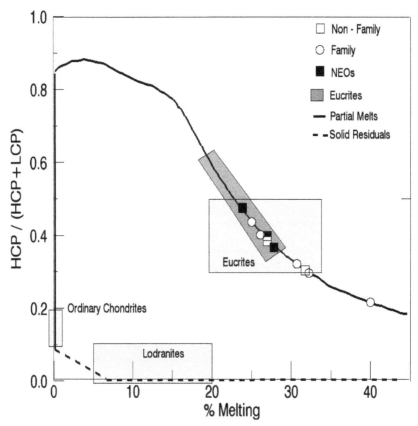

Fig. 12. High to Low calcium ratios in the solid residual and crystallized partial melt. Figure adapted from Sunshine et al. (2004)

In this work it was applied a spectral deconvolution technique that can separate end-members and determine the mineral composition present in the surface of an asteroid. As can be seen, the results are independent of the initial input. The starting point was a combination of 25/75 of HCP/LCP relation, and with a special mineralogical composition. From the sample of asteroids, it can be determined that 5 out of 11 (45%) from the Vesta family are Eucrite type, meanwhile 2 out of 6 (33%) non-family members are Eucrite type and also 100% of NEO. The small number of object in the sample needs to be taking into account here. Asteroid with presence of HCP (Eucrite type) are in all the three groups and HCP varies from 19 to 46%, which represent a melting between 15 to 35%.

The basaltic material outside the Vesta family is a nice and intriguing corundum. The Vesta family is defined as all the objects that can be ejected from collisions but allowed Vesta´s crust to survive. The limitation in collision energy or ejecta velocity is well restricted. Anyway, there are some basaltic objects outside the family, but in the region that can be explained by other alternative scenarios. They could be fragments from collisions on Vesta's crust and then transported to the current position (Carruba et al 2005). Or there are fragments from a different differentiated parent body, like Eunomia (Nathues et al. 2005) or Baptistina (Carvano and Lazzaro, 2010).

The only way to define without any doubt the mineralogy of the surface of an asteroid is going there. Several space mission visited many kind of asteroids, most of them doing only a fly-by. In the last decade, asteroids have become primary targets for space missions geared towards improving our understanding of SS formation. The Galileo mission, in 1991 (Russell, 1992), was the first to perform an asteroid flyby. On its way to Jupiter, it performed a flyby of the asteroids Gaspra and Ida, including its moon Dactyl. The NEAR-Shoemaker mission (Veverka et al., 2001) made a flyby of the asteroid Mathilde, and performed a rendezvous mission to Eros. Deep Space 1 (Nordholt et al., 2003) made a double fly-by of the asteroid Braille and comet Borelli. The Stardust mission (Brownlee et al., 2003) returned a sample of Comet Wild 2's coma material, and completed a flyby of asteroid Anne-Frank. In 2003, the Japanese Space Agency's (JAXA) Hayabusa mission was the first to focus on an asteroid with an aim to return a sample (Fujiwara et al., 2004). With only a basic scientific payload onboard Hayabusa, data gathered on the S-type asteroid Itokawa have nevertheless yielded incredible results (Nagao et al. 2011, Tsuchiyama et al. 2011) Ideas for visiting a V-type NEO in a sample return mission are presented in the work of Duffard et al. (2011).

4. DAWN mission at Vesta

It's really an exclusive opportunity to have a spacecraft orbiting the body that you are/were studying during years. This is the case for the Dawn mission and asteroid Vesta. Dawn was launched on 27 September 2007 and the mission's goal is to characterize the conditions and processes of the solar system's earliest eon by investigating in detail two of the largest proto-planets remaining intact since their formation. Ceres and Vesta have many contrasting characteristics that are thought to have resulted from them forming in two different regions of the early solar system. In a first stage of the journey, the spacecraft arrived at Vesta in August 2011 and will stay in two different orbits (low and high altitude) during one year after which will continue the journey to Ceres.

This spacecraft has an image-framing camera (FC), a visible and near infrared spectrometer (VIR) and a gamma ray and neutron detector (GRaND). The FC has a filter wheel with seven colour and one clear filter. An image consists of a frame of 1024 x 1024 pixels and one pixel has a field of view (FOV) of 93 μrad. The VIR mapping spectrometer is a compact spectrometer with both visible and infrared ranges: 0.25–1.0 and 0.95–5.0 μm. Its spatial resolution is 0.250 μrad with spectral resolution varying from 30 to 170. The GRaND instrument features neutron spectroscopy using Li-loaded glass and boron-loaded plastic phoswich. The gamma ray detection uses bismuth germanate and cadmium zinc telluride (Rusell, et al. 2006).

With all this instruments and after 1 year of orbiting Vesta the quantity and quality of information on this specific body will be incredible. As early as this chapter is written, first images on the surface of Vesta are arriving with unprecedented details [1].

During the early arrival to the asteroid the images could be compared with the one taken from the Hubble Space Telescope, showing that studies from ground are pretty close in details as studies from orbit. After entering the high altitude orbit, a mapping of the entire surface down to 260 meters per pixel was done. Unprecedented details on the surface can be observed. This is the first time a spacecraft is in orbit on an asteroid of this size and the feature characteristics are really unique. Thousands of new features are in the images, craters from several kilometres to some meters, grooves, faults, hills, etc. To name and characterize all these new features, a new reference system need to be taken. The zero-longitude, or prime meridian, of Vesta was defined by the science team using a tiny crater about 500 meters in diameter, which they named "Claudia," after a Roman woman during the second century B.C. Dawn's craters will be named after the vestal virgins – the priestesses of the goddess Vesta, and famous Roman women, while other features will be named for festivals and towns of that era.

One of the prominent details shown in the images are the equatorial ridges. Also there are grooves in the equatorial region of about 10 kilometres wide. Another distinct feature is a massive circular structure in the South Pole region. Scientists were particularly eager to see this area close-up. This feature is thought to be the origin of all the fragments forming the Vesta family and some V-type asteroids. Most of the HED meteorites would come from this impact region. The circular structure, or depression, is several hundreds of kilometres wide, with cliffs that are also several kilometres high. One impressive mountain in the centre of the depression rises approximately 15 kilometres above the base of this depression, making it one of the highest elevations on all known bodies with solid surfaces in the solar system.

As shown in Figure 13, the surface of Vesta is populated with craters in different sizes, showing a rather old surface. In some of the large craters it can be seen material movement to the centre due to gravity. Some small craters superimpose to the larger ones showing the intense bombardment that Vesta was affected in the past. Many of the smaller craters could also be part of the material ejected in the formation of the bigger craters. The grooves that are visible in the right part of the figure will give clues on the formation of the crust and discussion on the possible origin of this feature are open.

In next figure 14, it's shown that are some craters over the grooves which is indicative of the time of formation of the craters and grooves. Some worm-like features are also visible and there is still no explanation on the origin of this kind of feature. Dark and light material showing differences in compositions are also visible. Dark material can be organics, maybe rest of a cometary impact on Vesta. More detailed images on this region to try to explain the presence of this material will be taken in a lower orbit.

---

[1] All images in this section were taken with the Framing Camera. The framing cameras were developed and built under the leadership of the Max Planck Institute for Solar System Research, Katlenburg-Lindau, Germany, with significant contributions by the German Aerospace Center (DLR) Institute of Planetary Research, Berlin, and in coordination with the Institute of Computer and Communication Network Engineering, Braunschweig. The framing camera project is funded by NASA, the Max Planck Society and DLR. JPL is a division of the California Institute of Technology, in Pasadena.

The analysis, not only from the FC but also mainly from the VIR data will give us information on the different materials present in the craters. The South Pole region is the more interesting one because with this 15 km depth can show evidence of the upper mantle of Vesta. Several models were published discussing the probable depth of the crust, and the analysis of the images and spectra of this region will strongly constraint the models.

Fig. 13. This image was taken through the FC camera's clear filter and has a resolution of 260 meters per pixel. Image credits *NASA/ JPL-Caltech/ UCLA/ MPS/ DLR/ IDA*

In Figure 15 there is a comparison of the clear light picture and the false coloured one. Colours in the right picture are given to enhance differences in composition of the surface. In this false Red-Green-Blue (RGB) colour scheme, red is used for the ratio of the brightness at wavelengths of 750 nanometres to the brightness at 440 nanometres, green is used for the ratio to the brightness of 750 nanometres to 920 nanometres and blue is used for the ratio to the brightness at 440 nanometres to 750 nanometres. Red-blue tones capture the visible continuum and green tones capture the relative strength of the ferrous absorption band at 1 micron.

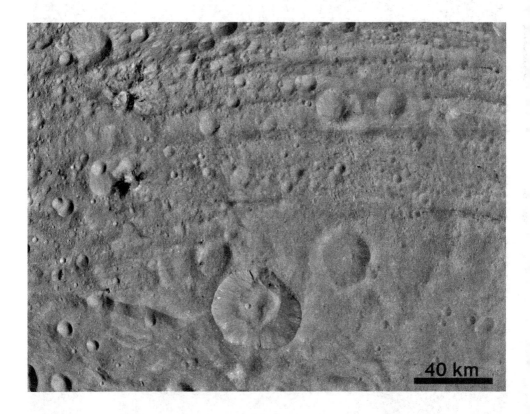

Fig. 14. This image, taken through the framing camera's clear filter, shows dark material at impact craters, up to 20 kilometre-wide and sets of worm-like tracks in the north-south direction. The image has a resolution of 254 meters per pixel. Image credits *NASA/ JPL-Caltech/ UCLA/ MPS/ DLR/ IDA*

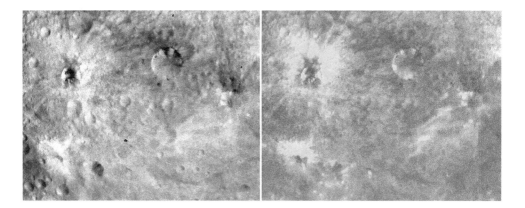

Fig. 15. Same region of Vesta taken in clear filter on false coloured images. The images have a resolution of about 490 meters per pixel. Image credits *NASA/ JPL-Caltech/ UCLA/ MPS/ DLR/ IDA*

5. Conclusions

The chapter started presenting the problem of the basaltic material on the minor bodies of our solar system. Then, two different techniques were presented to obtain clues in the mineralogy of these bodies. These methods showed that is possible to remotely obtain precise information on the surface of a body that can be visited by a spacecraft and confirm the findings. Data from those missions are powerful sources of information in details on the geology and mineralogy of the surfaces. After this entire journey the chapter can be ended with some conclusion and future work.

Vesta is NOT the only basaltic minor body in our solar system. Vesta is the only intact basaltic asteroid or basaltic proto-planet. Vesta has a "family" of objects that are related to them as to be the fragments of a collision or collisions. Some of these fragments are outside the boundary for some family determinations and these boundaries are related with the energy that a great impact would eject material from the crust but keeping the body as it can be observed today. One postulated option is that these fragments can be part of Vesta ejected material and then dynamically moved to their actual position outside the Vesta family. The other possibility is they are part of another differentiated parent body in the region. In the last years there are evidences that more differentiated bodies existed in the beginning of the solar system and now it can be observed fragments from the original ones. There is still no clear technique to remotely distinguish the mineralogy from one parent body to the other. Dynamics is needed to help in the solution of this problem. How is

possible to remotely indentify a fragment from deep in the crust-mantle of a specific differentiated parent body with another from other different parent body? Here several professional interact from different areas: dynamics, astrophysics, geophysics and mineralogy to help find the answers.

A problem to be solved is the apparent homogeneity in the known asteroid's families. Mineralogical studies of the asteroids families, shows that there is no evidence of a differentiated family. This mean, there is no family with fragments similar to the crust, fragments similar to the mantle and fragments similar to the nucleus of the parent body. All families seem to be the catastrophic evidence of homogeneous parent bodies. A question that can be raised here is if the collision that broke-up that differentiated objects were in the beginning of the solar system, in the first megayears, can we identify those catastrophic families? Differentiated objects formed in the first 5 megayears and have nearly 600 My to be disrupted before the Late Heavy Bombardment. If the differentiated families formed before the LHB it will not be possible to identify those families because fragments will be dispersed or ejected from the solar system.

On the other hand, in the meteorite collection there are some clear example and proofs that there were more than one differentiated asteroid. This proof came not only from the basaltic meteorites (as shown in Sections 2), but also from the iron meteorites, fragments of the nucleus of those differentiated asteroids. All these differentiated objects were formed in the first 4-5 mega-years of evolution of the solar system, otherwise the ^{26}Al is not enough powerful to melt and differentiate the body. It's expected that several dozens of differentiated bodies formed from 40 to 500 km diameter. Some of these were destroyed by catastrophic impacts; some others formed bigger objects, some other, at least one, survived. Our Earth took 30 mega-years to be formed so its formation was with proto-planets that were differentiated and more primitive material. The same occurred with Mercury, Venus and Mars with different formation time depending on size.

Vesta is the only intact basaltic body. Here the word "intact" is intend to mention that Vesta has the basaltic crust and remains almost as it formed in the beginning of the solar system. As can be seen in the Dawn images of the preceding section, Vesta's crust is far to be "intact". An interesting part of the crust of this body is the South Pole region where a giant impact consequence can be seen. Most of the material ejected from this impact can be seen today as members of the Vesta dynamical family. Studying these fragments is a way to go deep in the crust and this is the main reason to study the mineralogy of family and non-family members and also V-type NEO's.

In this work I presented the results of a new analysis on the V-type asteroids. The mineralogy and LCP to HCP ratios of those V-type asteroids were analysed and compared to the HED meteorite using two different techniques. First, the two methods were applied to the HED sample to gain practise and be confidant and then applied to the asteroids where the quality of the data is lower. Combining all the data from this work and from literature a nice scenario of what happened in the crust of Vesta is revealed. Large impact/s excavated material from the crust and studying these fragments its possible to obtain information on different depths of the crust and maybe the mantle. From the current study (and others from the literature) its possible to infer the mineralogy of these fragments. Some of them are

Eucrite type showing a composition with LCP and HCP in different proportions. Other objects show a Diogenite composition testing deeper layers in the crust. Results on the calcium and iron content of the pyroxene were obtained here and in other works indicating that conclusion.

The studied V-type asteroids have a large range of mineralogy but with a prevalence of asteroids similar to Diogenites type material. This result goes in the same direction of Vernazza et al. (2005), suggesting that the spectral diversity observed in the Vestoids arises because these asteroids contain material present on the Vesta surface or coming from different layers of Vesta excavated by collisions and not because they originated from different parent bodies. After a big collision, fragment from the crust or re-accumulated fragments should show Howardite type spectra. This kind of spectra can be characteristics of asteroids where both Eucrite and Diogenite materials are present. A re-accumulated fragment after a collision should fulfil this requirement. As it can be seen in the previous section, the surface of Vesta is inhomogeneous, with lots of impacts that excavated material from deeper layers. A giant impact would extract material from different depth and then fragments would re-accumulate to form a Howardite type family member asteroid.

There are also some objects that can show evidence of olivine. If the collision that formed the South Polar Region excavate enough material to reach the mantle, it will be possible to identify olivine in the fragments. This assumption needs to be confirmed. Current magma ocean models do not predict heterogeneity at the small kilometre scale of the V-type asteroids, on the other hand modellers seems to support the idea that partial melting may have played a important role in the formation of Vesta.

Finally, the mineralogical composition and internal structure of the primitive Earth is unknown. Plate tectonic, mountain creation and erosion destroyed the geological information from that period. An innovative way to understand all this primitive process is to compare the Earth with bodies that did not suffer large alterations, like Mars, the Moon, Mercury or differentiated asteroids. Primitive Earth formed from the accumulations of planetary embryos, some of them already differentiated, with different sizes (Allegre et al. 1995). The most important of those bodies was a Mars sized differentiated proto-planet that collides with the still forming Earth. From this catastrophic event, the Moon formed (Zhang 2002). To understand the formation of the primitive Earth is important to know how those differentiated bodies formed. Bigger the body, larger was the time it took to accrete in a proto-planet (Kleine et al. 2002). At least, a large number of the bodies that accreted in the terrestrial planets were differentiated.

All this research is based in the interpretation of surface reflectance spectra obtained in ground-based telescope, space telescope and laboratories (known minerals or meteorites). Next step is to include the mineralogy of the exo-planets that are being discovered in the last years. The so called exo-Earths are starting to be discovered in the last 3 years and theories recall to be originated in at least two different ways: the formation of a big terrestrial planet with 3-5 times the mass of the Earth or they are the bared nucleus of a giant planet that migrated closer to the star and all the gas envelope were flown away.

Now the question to be answered is which is the mineralogy of these planets? In our case, the less altered meteorites have almost the same composition of our star, the Sun. They

formed from the same primordial nebula, so with the exception of the hydrogen and helium that are mainly in the central star; all the mineral proportion should be the same. In the near future, when systems with several planets will be identified, and several Earth like planets be present on those system the possible composition of those planets can be inferred from the composition of the star. The chemistry of the central star will have a clue on the material that formed these planets.

It's curious, to know the mineralogy of a planet, some introspective looking, we also need to study the star that hosts this planet.

6. Acknowledgment

I want to thank to the financial support obtained in my current Ramón y Cajal contract from the Ministerio de Ciencia e Innovación of Spain.

7. References

Adams, J.B. (1974). Visible and near-infrared diffuse reflectance spectra of pyroxenes as applied to remote sensing of solid objects in the Solar System. *J. Geophys. Res.* 79, 4829–4836.

Adams, J.B. (1975). Interpretation of visible and near-infrared diffuse reflectance spectra of pyroxenes and other rock-forming minerals. In: Karr Jr., C. (Ed.), *Infrared and Raman Spectroscopy of Lunar and Terrestrial Minerals*. Academic Press, San Diego, pp. 91–116.

Allegre, C., Manhes, G. and Gopel, C. (1995). The age of the Earth. *Geochimica et Cosmochimica Acta, Vol.* 59, pp. 1445-1456.

Asimov, P.D., M.S. Ghiroso (1998). Algorithmic modifications extending MELTS to calculate sub-solidus phase relations. *Am. Mineral.* 83, 1127

Binzel, R.P., Xu, S., (1993). Chips off of Asteroid 4 Vesta - evidence for the parent body of basaltic achondrite meteorites. *Science* 260, 186–191.

Binzel, R.P.; Birlan, M.; Bus, S.J.; Harris, A.W.; Rivkin, A.S.; Fornasier, S. (2004). Spectral observations for near-Earth objects including potential target 4660 Nereus : Results from Meudon remote observations at the NASA Infrared Telescope Facility (IRTF). Planetary and Space Science, Volume 52, Issue 4, p. 291-296.

Bland, P. et al. (2009). An Anomalous Basaltic Meteorite from the Innermost Main Belt. *Science*, Volume 325, Issue 5947, pp. 1525-

Brownlee, D.E., Tsou, P., Anderson, J.D., Hanner, M.S., Newburn, R.L., Sekanina, Z., Clark, B.C., Horz, F., Zolensky, M.E., Kissel, J., et al. (2003) Stardust: comet and interstellar dust sample return mission. *Journal of Geophysical Research* 108 (E10), 1–15.

Bottke, W.F. et al., (2006). Iron meteorites as remnants of planetesimals formed in the terrestrial planet region. *Nature* 439, 821–824

Burbine, T.H., Buchanan, P.C., Binzel, R.P., Bus, S.J., Hiroi, T., Hinrichs, J.L.,Meibom, A., McCoy, T.J., (2001). Vesta, vestoids, and the howardite, eucrite, diogenite group: relationships and the origin of spectral differences. *Meteorit. Planet. Sci.* 36, 761–781.

Burns, R.G., (1970). Crystal field spectra and evidence of cation ordering in olivine minerals. *Am. Mineral.* 55, 1608–1632.

Canas,L., Duffard, R, Seixas, T. (2008). Mineralogy of HED meteorites using the modified Gaussian model. *Earth Moon Planets* 102(1–4), 543–548

Carruba, V.; Michtchenko, T. A.; Roig, F.; Ferraz-Mello, S.; Nesvorný, D. (2005). On the V-type asteroids outside the Vesta family. I. Interplay of nonlinear secular resonances and the Yarkovsky effect: the cases of 956 Elisa and 809 Lundia. *Astronomy and Astrophysics*, Volume 441, Issue 2, pp.819-829

Clayton, R. N.; Mayeda, T. K. (1983). Oxygen isotopes in eucrites, shergottites, nakhlites, and chassignites. *Earth and Planetary Science Letters*, vol. 62, no. 1, p. 1-6.

Clayton, R. N.; Mayeda, T.K. (1996). Oxygen isotope studies of achondrites. *Geochimica et Cosmochimica Acta*, vol. 60, Issue 11, pp.1999-2017.

Clayton, D. D.; Meyer, B.S.; The, L.; El E., Mounib F. (2002). Iron Implantation in Presolar Supernova Grains. *The Astrophysical Journal,* Volume 578, Issue 1, pp. L83-L86.

Cloutis, E.A., (1985). Interpretive techniques for reflectance spectra of mafic silicates. *MSc thesis. Univ. of Hawaii, Honolulu.*

Cloutis, E.A., Gaffey, M.J., (1991). Pyroxene spectroscopy revisited spectral-compositional correlations and relationship to geothermometry. *J. Geophys. Res.* 96 (E5), 22809–22826.

Cruikshank, D.P., Tholen, D.J., Bell, J.F., et al., (1991). Three basaltic earth-approaching asteroids and the source of the basaltic meteorites. *Icarus* 89, 1.

de León, J.; Licandro, J.; Duffard, R.; Serra-Ricart, M. (2006) Spectral analysis and mineralogical characterization of 11 olivine pyroxene rich NEAs. *Advances in Space Research,* Volume 37, Issue 1, p. 178-183.

de Sanctis, M. C.; Migliorini, A.; Luzia Jasmin, F.; Lazzaro, D.; Filacchione, G.; Marchi, S.; Ammannito, E.; Capria, M. T. (2011). Spectral and mineralogical characterization of inner main-belt V-type asteroids. *Astronomy & Astrophysics*, Volume 533, id.A77.

Duffard, R.; Lazzaro, D.; Licandro, J.; deSanctis, M. C.; Capria, M. T.; Carvano, J.M. (2004). Mineralogical characterization of some basaltic asteroids in the neighborhood of (4) Vesta: first results. *Icarus,* Volume 171, Issue 1, p. 120-132.

Duffard, R.; Lazzaro, D.; de León, J. (2005). Revisiting spectral parameters of silicate-bearing meteorites. *Meteoritics & Planetary Science,* Vol. 40, p.445

Duffard, R., deLeón, J., Licandro,J., Lazzaro,D., Serra-Ricart,M., (2006). Basaltic asteroids in the near-Earth objects population: a mineralogical analysis. *Astron. Astrophys.* 456,775.

Duffard, R.; Roig, F. (2009) Two new V-type asteroids in the outer Main Belt? *Planetary and Space Science,* Volume 57, Issue 2, p. 229-234.

Duffard, R.; Kumar, K.; Pirrotta, S.; Salatti, M.; Kubínyi, M.; et al. (2011). A multiple-rendezvous, sample-return mission to two near-Earth asteroids. Advances in Space Research, 48, p. 120-132.

Drake, M.J., (2001). The eucrite—Vesta story. *Meteorit. Planet. Sci.* 36, 501– 513.

Florczak, M., Lazzaro, D., Duffard, R., (2002). Discovering new V-type asteroids in the vicinity of 4 Vesta. *Icarus* 159, 178–182.

Fujiwara, A., Kawaguchi, J., Uesugi, K.T. Role of sample return misión MUSES-C in asteroid study.(2004). *Advances in Space Research* 34, 2267– 2269.

Gaffey M. J., Cloutis E. A., Kelley M. S., and Reed K. L. (2002). Mineralogy of asteroids. In *Asteroids III*, edited by Bottke W. F. Jr., Cellino A., Paolicchi P., and Binzel R. P. Tucson, Arizona: The University of Arizona Press. pp. 183–204.

Ghiorso, M.S.; Sack, R.O. (1995). Chemical mass transfer in magmatic processes IV. A revised and internally consistent thermodynamic model for the interpolation and extrapolation of liquid-solid equilibria in magmatic systems at elevated temperatures and pressures. *Contributions to Mineralogy and Petrology*, Volume 119, Issue 2/3, pp. 197-212.

Gupta, G. and Sahijpal S. (2010). Differentiation of Vesta and the parent bodies of other achondrites. *Journal of Geophysical Research* 115, E080001.

Ivezic, Z ., Tabachnik, S., Rafikov, R.,et al., (2001). Solar system objects observed in the Sloan Digital Sky Survey commissioning data. *Astron. J.* 122,2749.

Juric, M., Ivezic, Z., Lupton, R.H., et al., (2002). Comparison of positions and magnitudes of asteroids observed in the Sloan Digital Sky Survey with those predicted for known asteroids. *Astron.J.* 124,1776.

King, T. V. V.; Ridley, W. I. (1987). Relation of the spectroscopic reflectance of olivine to mineral chemistry and some remote sensing implications. *Journal of Geophysical Research* (ISSN 0148-0227), vol. 92, Oct. 10, p. 11457-11469.

Kleine, T. , C. Münker, K. Mezger, H. Palme. (2002). Rapid accretion and early core formation on asteroids and the terrestrial planets from Hf-W chronometry. *Nature* 418(6901), 952–955

Lazzaro,D., T.A. Michtchenko, J.M. Carvano et al., (2000). Discovery of a basaltic asteroid in the outer Main Belt. *Science* 288, 2033–2035

Carvano, J. M.; Lazzaro, D. (2010). Diameter, geometric albedo and compositional constraints for (298) Baptistina through visible and mid-infrared photometry. *Monthly Notices of the Royal Astronomical Society: Letters*, Volume 404, Issue 1, pp. L31-L34.

Mayne, R. G.; Gale, A.; McCoy, T. J.; McSween, H. Y., Jr.; Sunshine, J. M. (2006). The Unbrecciated Eucrites: Vesta's Complex Crust. *Meteoritics & Planetary Science*, Vol. 41, Supplement, Proceedings of 69th Annual Meeting of the Meteoritical Society, held in Zurich, Switzerland., p.5093.

McFadden, L., Gaffey, M.J., McCord, T., (1985). Near-earth asteroids— possible sources from reflectance spectroscopy. *Science* 229, 160.

Mittlefehldt, D. W.; Lindstrom, M. M. (1998). Petrology and Geochemistry of Lodranite GRA 95209. *Meteoritics & Planetary Science*, vol. 33, p. A111

Moskovitz, Nicholas A.; Jedicke, Robert; Gaidos, Eric; Willman, Mark; Nesvorný, David; Fevig, Ronald; Ivezić, Željko. (2008) The distribution of basaltic asteroids in the Main Belt. *Icarus*, 198, pp. 77-90.

Nagao, K. Et al. (2011). Irradiation History of Itokawa Regolith Material Deduced from Noble Gases in the Hayabusa Samples. *Science,* Volume 333, Issue 6046, pp. 1128-

Nathues, A.; Mottola, S.; Kaasalainen, M.; Neukum, G. (2005).Spectral study of the Eunomia asteroid family. I. Eunomia. *Icarus,* Volume 175, Issue 2, p. 452-463.

Nordholt, J.E., Reisenfeld, D.B., Wiens, R.C., Gary, S.P., Crary, F., Delapp, D.M., Elphic, R.C., Funsten, H.O., Hanley, J.J., Lawrence, D.J., et al. (2001) Deep Space 1 encounter with Comet 19P/Borrelly: ion composition measurements by the PEPE mass spectrometer. *Geophysical Research Letters* 30 (9), 18–21.

Pieters, C.M., Hiroi, T., (2004). RELAB (Reflectance Experiment Laboratory): A NASA multi-user spectroscopy facility. *Lunar Planet. Sci.* 35. Abstract #1720 (CDROM).

Pieters, C.M., R. P. Binzel, D. Bogard, T. Hiroi, D.W. Mittlefehldt, L. Nyquist, A. Rivkin and H. Takeda. (2005). Asteroid-meteorite links: the Vesta conundrum(s). *Asteroids, Comets, Meteors Proceedings IAU Symposium No. 229, D. Lazzaro, S. Ferraz-Mello & J.A. Fernández, eds.*

Russell, C.T.(1992) The Galileo mission. *Space Science Reviews* 60 (1-4/ CONF.), 1–2.

Russell, C.T, at al. (2006) Dawn discovery mission to Vesta and Ceres: Present status. *Advances in Spce Research,* 38, pp: 2043-2048.

Sunshine, J., Pieters, C., Pratt, S., (1990). Deconvolution of mineral absorption bands: an improved approach. *J. Geophys. Res.* 95 (B5), 6955– 6966.

Sunshine, J.M. , C.M. Pieters (1993). Estimating modal abundances from the spectra of natural and laboratory pyroxene mixtures using the modified Gaussian model. *J. Geophys. Res.* 98(E5), 9075–9087

Sunshine, J.M. S.J. Bus, T.J. McCoy, T.H. Burbine, C.M. Corrigan, R.P. Binzel. (2004). High-calcium pyroxene as an indicator of igneous differentiation in asteroids and meteorites. *Meteoritics & Planet. Sci.* 39, 1343–1357

Takeda, H., (1997). Mineralogical records of early planetary processes on the HED parent body with reference to Vesta. *Meteorit. Planet. Sci.* 32, 841–853.

Thomas, P.C., Binzel, R.P., Gaffey, M.J., et al., (1997). Impact excavation on asteroid 4 Vesta: Hubble Space Telescope results. *Science* 277, 1492.

Tsuchiyama, A. et al. (2011). Three-Dimensional Structure of Hayabusa Samples: Origin and Evolution of Itokawa Regolith. *Science,* Volume 333, Issue 6046, pp. 1125-

Vernazza, P.; Mothé-Diniz, T.; Barucci, M. A.; Birlan, M.; Carvano, J. M.; Strazzulla, G.; Fulchignoni, M.; Migliorini, A. (2005). Analysis of near-IR spectra of 1 Ceres and 4 Vesta, targets of the Dawn misión. *Astronomy and Astrophysics,* Volume 436, Issue 3, pp.1113-1121.

Veverka, J., Farquhar, B., Robinson, M., Thomas, P., Murchie, S., Harch, A., Antreasian, P.G., Chesley, S.R., Miller, J.K., Owen, W.M., et al. (2001). The landing of the NEAR-shoemaker spacecraft on asteroid 433 Eros. *Nature 413* (6854), 390–393.

Williams, J.G., (1989). Asteroid family identifications and proper elements. In: Binzel, R.P., Gehrels, T., Matthews, M.S. (Eds.), *Asteroids II.* Univ. of Arizona Press, Tucson, pp. 1034–1072.

Yamaguchi, A., Clayton, R.N., Mayeda, T.K., et al., (2002). A new source of basaltic meteorites inferred from Northwest Africa 011. *Science* 296, 334.

Zhang, Y. (2002). The Age and accretion of the Earth. *Earth-Science Reviews* 59, pp 235-263.

Zappalá, V., Cellino, A., Farinella, P., Knezevic, Z., (1990). Asteroid families. I. Identification by hierarchical clustering and reliability assessment. *Astron. J.* 100, 2030–2046.

Pathways for Quantitative Analysis by X-Ray Diffraction

J.D. Martín-Ramos[1], J.L. Díaz-Hernández[2],
A. Cambeses[1], J.H. Scarrow[1] and A. López-Galindo[3]
[1]University of Granada, Department of Mineralogy and Petrology
[2]IFAPA Camino de Purchil, Departament of Natural Resources, Junta de Andalucía
[3]Instituto Andaluz de Ciencias de la Tierra (IACT, CSIC-UGR)
Spain

1. Introduction

Quantitative analysis by powder X-Ray diffraction (Q-PXRD) is hardly ever carried out by simple or obvious methods. Although the theory has been well established for some time (Clark, 1936; Alexander et al., 1948; Chung, 1974, 1975; Brindley, 1980; Bish et al., 1988; etc.), in practice imprecise results are obtained because of experimental oversimplification usually introduced as a result of the complexity of the problem. Most of the usual methods are based on the correspondence between the intensities diffracted by each crystalline phase and their weight content in a multiple component mixture, without previously assuming that this correspondence is linear.

A straightforward simplification of the general equation of the intensity diffracted by a crystal [1] allows us to estimate the weight fraction (X_p) of a phase (P) diffracting at intensity $I_{hkl,P}$[2]:

$$I_{hkl,P} = I_0\lambda^3/64\pi r \cdot e^2/m_e c^2 \ W_{hkl}/V^2{}_P \cdot |F_{hkl,P}|^2 \cdot Lp_\theta \cdot X_{v,P}/\mu_S \qquad (1)$$

(W_{hkl} = multiplicity, V = unit-cell volume, Lp_θ = Lorentz-polarization factor, $X_{v,P}$ = volume fraction of phase P)

$$X_P = K_D \cdot (\mu/\rho)_S \cdot (M \cdot \rho \ [C \cdot I/|F|^2]_{hkl})_P \qquad (2)$$

In this expression, the subscripts refer to each phase (P) of the sample (S), to the experimental geometry of the diffractometer (D) and to the reflections (*hkl*) of each of the individual phases. This means that analysis quality is conditioned by factors affecting not only each component of the sample analyzed, but also other factor including the set of minerals making up the sample $(\mu/\rho)_S$, texture ($C_{hkl,P}$) and the experimental device itself K_D, the last being a general constant considered to be unvarying in all diffraction experiments in a given laboratory.

2. Methods of quantitative analysis

The various ways in which the parameters of formula [2] can be simplified give rise to the various methods of Q-PXRD.

2.1 Semi-quantitative or simple quantitative analysis

In this case, a single reflection is used per phase, absorption correction is omitted and the result is adjusted so that the sum of $X_P = 1$. The results obtained by this method usually have considerable error and are not susceptible to statistical analysis. In other words:

$$X_P = ([M \cdot I_{DBmax}/I_{DB} \cdot I]_{hkl})_P \tag{3}$$

with $M_{hkl,P}$ being equivalent to $1/RIR$, I_{DBmax} = Maximum intensity of standard database patterns (normally 100 or 1000) and I_{DB} = Intensity of used reflection from database patterns. When the maximum intensity reflection is used, the value of I_{DBmax}/I_{DB} is obviously 1.

The example shown in table 1 corresponds to a semi-quantitive analysis of the diffractogram from figure 1 (sample F3). The scale factors are taken from database PDF2 (International Centre for Diffraction Data, 2003), the reflections used were those with the highest intensities and without interferences with other phases. Percentages are given as whole numbers given that the maximum permitted precision is around 10%. The correct results are those given in table 3 (see below) they differ significantly from the results in table 1.

Phase	hkl	Set-file	Int(100)	RIR	%
Aragonite	111	41-1475	7.4	1.00	12
Calcite	104	81-2027	100.0	0.31	49
Celestine	121	89-7355	10.2	0.53	8
Dolomite	104	73-2324	50.0	0.40	31
Sum					100

Table 1. Simple quantitative analysis (semi-quantitative) of a four high crystallinity components sample. Reference intensity ratio (RIR) are from PDF2 database pattern I/Icor values.

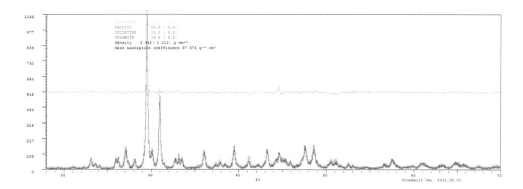

Fig. 1. Diffractogram pattern of the mineral mixture indicated. Grey points=calculated pattern. Purple line=observed pattern. Green = Difference between observed and calculated diffractograms + 50 % elevation.

In a general way, by adding a known amount of internal-standard to the sample we can substantially improve the results of the semi-quantitative method. There is an abundance of bibliographical references on this subject beginning with Chung (1974). Otherwise, simplification is achieved by using the relation between the mass coefficients of the total sample and of each component (Absorption-Diffraction method).

$$X_P = (I_{hkl,S} / I_{hkl,P}) \cdot (\mu/\rho)_P / (\mu/\rho)_S \tag{4}$$

2.2 Rietveld methods

In this case the complete profile of the theoretical diffractogram that best fits the experimental results is calculated on the basis of the known crystalline structures of the sample components, the fit of some instrumental parameters, and the distribution functions describe the shape of the reflections.

The method depends on the relationship:

$$W_P = (S \cdot M \cdot V)_P / \sum (S \cdot Z \cdot M \cdot V)_i \tag{5}$$

(W = weight fraction, P = each phase, n = number of phases, i = blank notation (1 to n), S = Scale factor, Z = Number of formulas per unit-cell, M = Molecular weight, V = unit-cell Volume).

As in other quantitive methods it is necessary to use adequate Reference Intensity Ratio (RIR) to obtain the best results. Non-linear minimal squares methods are used that minimize the accordance factor, R, whose most general form is:

$$R = \sum [w \mid I_o - I_c \mid]_i / \sum [w \, I_o]_i \tag{6}$$

(i = each experimental point (blank, 1 to number of measured points), o = observed intensities, c = calculated intensities, w = statistic weight). The quantitative results are obtained on the basis of the scale factors used by each diffractogram partially determined by the experimental conditions. This process refines: the unit cell, coordinates, factors of temperature and atomic occupancies, profile parameters (mainly width and asymmetry), 2θ displacements, preferred orientation, background radiation parameters, extinction and microabsorption. The orientation factors and scale factors of each phase are also refined. This method is elegant. The R values achieved can be very low and the diffractograms of differences $(I_o-I_c)_i$ are usually very flat (i is each point of the diffractogram), although at times the quality of an analysis may be masked by the complicated mathematical operations and the results can have an 'excessively' theoretical component. There are there are several excellent programs that use Rietveld methods in general, and some have been specifically created for use in quantitative analysis (Quanto). See http://www.ccp14.ac.uk/solution/rietveld_software/index.html. Typical Rietveld analyses for sample F3 are shown in table 2 and in figure 2.

2.3 Full pattern fitting using experimental data

The relative mixtures of experimental diffractograms of pure phases can be used to simulate a problem sample using least squares methods (LS) as in the Rietveld method (Rietveld, 1969). If the chemical composition and the cell parameters of each phase are known,

Phase	%	σ
Aragonite	12.9	0.3
Calcite	53.9	0.3
Celestine	12.2	0.2
Dolomite	21.0	0.2

Table 2. Quantitative Rietveld analysis. Experimental data: 2θ step 0.01 °; Radiation Cu Kα; Measured points= 5201; Counts per 8 sec; Total elapsed time = 702 min ; Maximum measured counts= 1247002 (divided by a suitable factor by set in appropriate scale); Final R_{Int} = 0.0382.

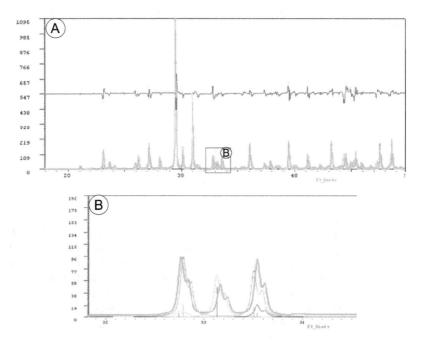

Fig. 2. A) Rietveld plot showing experimental patterns (green), calculated pattern (grey points), difference pattern (red line at diffractogram middle maximum). B) Details of Rietveld analysis of sample F3, Bragg´s bar positions and their unique full-profiles. 2θ positions of partial components are displayed before zero shift refinement

absorption corrections can be made. This is a real alternative to the Rietveld methods that can sometimes provide more realistic solutions. In general, we can omit the tuning of some instrumental factors and other aspects concerning the sample itself that might affect the final results, such as profile parameters, orientation parameters or radiation background. The method behaves like a black-box model regarding the crystalline structure of each phase, but this can only be an advantage when starting with phases with a chemical composition, crystallinity and orientation model similar to those of the sample in question. This method is

that of the *XPowder©* program, which uses a definition of the R_{Int} parameter very similar to the general one in equation [6].

$$R_{Int} = \{ \Sigma \left[w \left(I_o - I_c \right)^2 \right]_i \; / \; \Sigma \left[w \cdot I_o^2 \right]_i \} \, (n \cdot p) \tag{7}$$

where I_o are observed intensities and I_c calculated intensities, I = each experimental point, n = number of experimental points, p = number of standard patterns, $(n \cdot p)$ is constant for each analysis and is used only for scale purposes.

Another factor that has to be taken into account is the variation in absorption coefficients of each multiple component mixture and each component in isomorphic phases. A range of correction models have been used (for example see [4]), these usually produce different results that depend on the geometry of the diffractometer. The best result is usually obtained by using a multicomponent standard that has similar mass absorption coefficients to the analyzed sample. In this case the composition of each component is affected by a scale factor given by:

$$S_i = [X_i / X_p] \cdot [W\%_p / W\%_i] \tag{8}$$

where i is each component; P is the reference component that should be the most common in the samples (although not necessarily present in all of them); X is the weight fraction of each component of the diffractogram, and W% is the real weight of each component in the matrix W (see Appendix [A6]). Of course, these factors only have to be calculated for the standard sample, formed of the components i and P. This approximation gives sufficiently exact results.

An additional advantage is that amorphous phases can be treated in the same way as the crystalline compounds, if the isolated amorphous component is available. When all the partial diffractograms and the given sample are recorded in the same experimental conditions (same sample carrier, recording conditions, etc.) and the intensity of the incident rays is stable, it is not usually necessary to carry out any scale correction when absolute units or time units are used. However, if the intensities are used in relative mode (e.g., when the values are adjusted to a maximum value of 1000) a standard sample of known composition must be used to calculate the M_P factor of each phase.

This method also has the advantage that it is not necessary to perform the diffractogram analyses with the precision needed by the Rietveld method. As is well known the latter requires static analysis of various seconds per point (see 2.2, table 2), with between measurement intervals (2θ step) of 0.02 and 0.002° of 2θ. Analysis of a sample under such conditions can take up to a whole day. With the method discussed in this section, it is enough to use the typical working method of any diffractometer laboratory (12 to 30 minutes per analysis) but making sure to analyze standards under the same conditions. On the other hand, with the current method it is not necessary to adjust the 'instrumental function', 'Caglioti function', asymmetry, sample profile function, etc., given that these data are included in the standard analyses. In any case, it is clear that the results are more correct when the quality of the experimental data is better.

The main difficulty for application of this method is finding suitable standard substances. However, they can be obtained and when available the results are very precise. Table 3 shows the data for *XPowder©* program quantitative batch mode analysis of various samples that have the same qualitative composition as before.

Sample	Ara	Ara-σ	Cal	Cal-σ	Cel	Cel-σ	Dol	Dol-σ	R_{int}	Cycles	Sigma
F1	19.5	0.2	40.0	0.2	19.8	0.2	20.7	0.2	0.00499	3	0.00064
F2	15.6	0.2	33.6	0.2	32.8	0.2	18.0	0.3	0.00467	3	0.00060
F3	14.3	0.2	53.8	0.2	13.2	0.2	18.6	0.2	0.00369	3	0.00061
F4	27.6	0.3	21.5	0.4	41.0	0.3	10.0	0.5	0.00604	3	0.00068
F5	22.0	0.2	30.9	0.2	6.0	0.2	41.1	0.2	0.00439	2	0.00064

Table 3. Qualitative and quantitative composition of studied samples and standard deviation. Note that R values are <<0.01, indicating good adjustment. Known % weigh quantitative composition of sample F1 are: Ara: Aragonite 19.53, Cal: Calcite 39.91, Cel: Celestine 19.86 and Dol: Dolomite 20.71 was used as P component in [8]. With σ, as a standard deviation results of adjustment of each mineral.

In any case, there are multiple occasions when it is not necessary to evaluate the whole sample and so it is not necessary to have all the individual standards. The possibility to use natural or synthetic standards provides the opportunity to analyze individual components instead of a crystalline or partially amorphous matrix, always when a stable diffractogram is produced. As an example, it is possible to estimate the amount of a mineral in an organic matrix mix; the principal active pharmaceutical component amongst commercial additive such as starch or lactose; the amount of gypsum added to 'clinker' powder in a 'Portland' cement.

The given examples, and also the following, were obtained the *XPowder©* program that allows the component total to be adjusted to 100%, or alternatively to use one of the components, artificially added or not, as an internal standard with an aim to determine the absolute quantity of each independently of the composition of the components in the mix. The details of the calculation performed are in the appendix. The functions defined in the present work include adjustments of possible 2θ angle displacements, corrections related to component absorption coefficients as estimated from the database chemical formulas and absorption mass coefficient of the whole sample that is obtained iteratively from the calculated mineral composition. An option is that the intensity values can be statistically weighed with $I^{-1/2}$.

The application of this method is simple in transformations that convert one crystalline phase into another during a natural industrial process. It is also useful in crystalline processes in which the independent variable is, for example, time, concentration, pressure or temperature. The method is very precise when the chemical composition varies bit by bit, such as in dehydration for example. Crystallization of adhesives has been successfully tested as has crystallization of cements and cement glues and sulphate dehydration of Mg, Ca and Na (Cardell et al., 2007).

The following example (figure 3) is of the dehydration transformation of hidrotalcite. The process was monitored by X-ray diffraction and the results represented on a bidimensional map where colours represent relative intensities (greys represent low intensities). Angles are shown on the x-axis and temperatures on the y-axis. Recrystallization began during heating at around 100°C. In this example the absorption correction was not done because of lack of information about the samples true composition. Nevertheless, this correction should not be important given that few changes are noted in the global chemistry. For the quantitative calculation the first and last diffractograms were used as standards, the others were defined as a sample of these.

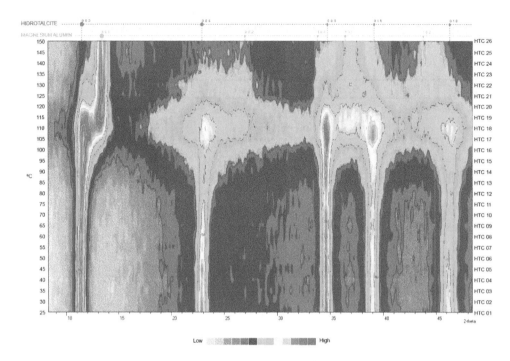

Fig. 3. *XPowder©* false colour 2D representation that shows the evolution of the diffractograms of hidrotalcite obtained during heating between 25-150°C. The image corresponds to the computer screen, so the characters are reduced to be able to read them in the present format: the header presents the database mineral information, in different colours and with diameters for each reflection point, including the *hkl* reflection. The false colour represents the relative intensities, absolute values are shown by contour lines.

Table 4 shows the low values obtained for R_{Int} and standard deviation (σ). This indicates the high quality of the results. These highest indices values correspond to temperatures between 95 and 115 °C. This is due to the presence of an important amorphous phase produced between the melting of the low temperature phase and the crystallization of the new high temperature phase. These results could be improved with a longer diffractogram analysis with a more gradual temperature increase. Diffractograms may also be used to 100 °C to quantify the process, such that an third amorphous low crystallinity component could easily be evaluated.

The method that adjusts the whole profile permits weighting between components that have an identical nature but somewhat different characteristics. For example, it is possible to use standards with very different crystallinities to weight the crystallinities of minerals intermediate compositions. This method has recently been used in very low crystallinity such as starch in the food industry. Use of this peculiarity is common in classic mineralogical studies about the measurement of illite, graphite, carbonates, etc. Despite this

it is common to weight the crystallinity using the classic method full width, half maximum (FWHM) that is not very sensitive in most cases, or is affected by the diffractometer optics especially by instrumental function which makes the analysis unreliable despite a frequent use of phase standards.

Sample	°C	Deh	σ	Hyd	σ	R$_{int}$	cycles	σ
HTC_001	25	0.0	0.0	100.0	0.0	0.00000	2	0.00000
HTC_002	30	1.7	0.6	98.3	0.5	0.00155	2	0.00005
HTC_003	35	1.5	0.6	98.5	0.5	0.00171	2	0.00006
HTC_004	40	1.7	0.6	98.3	0.5	0.00163	2	0.00005
HTC_005	45	1.8	0.6	98.2	0.5	0.00175	2	0.00006
HTC_006	50	3.1	0.7	96.9	0.5	0.00218	2	0.00006
HTC_007	55	3.7	0.7	96.3	0.6	0.00242	2	0.00007
HTC_008	60	4.8	0.8	95.2	0.6	0.00280	2	0.00007
HTC_009	65	6.6	0.9	93.4	0.7	0.00433	2	0.00009
HTC_010	70	8.6	0.9	91.4	0.8	0.00446	2	0.00009
HTC_011	75	12.4	1.0	87.6	0.9	0.00569	3	0.00010
HTC_012	80	15.2	1.1	84.8	1.0	0.00704	3	0.00012
HTC_013	85	21.3	1.4	78.7	1.2	0.00844	3	0.00013
HTC_014	90	25.9	1.5	74.1	1.4	0.00973	3	0.00014
HTC_015	95	33.3	1.8	66.7	1.6	0.01166	3	0.00016
HTC_016	100	41.2	1.8	58.8	1.7	0.01116	3	0.00016
HTC_017	105	50.0	2.1	50.0	2.1	0.01094	3	0.00016
HTC_018	110	59.2	1.8	40.8	1.8	0.00712	3	0.00013
HTC_019	115	67.5	1.5	32.5	1.6	0.00620	3	0.00012
HTC_020	120	74.6	1.1	25.4	1.1	0.00416	3	0.00010
HTC_021	135	79.9	0.8	20.1	0.9	0.00333	3	0.00009
HTC_022	140	85.1	0.7	14.9	0.8	0.00278	2	0.00008
HTC_023	145	90.9	0.6	9.1	0.7	0.00257	2	0.00008
HTC_024	150	95.3	0.5	4.7	0.6	0.00199	2	0.00006
HTC_025	155	97.3	0.5	2.7	0.6	0.00191	2	0.00006
HTC_026	160	100.0	0.0	0.0	0.0	0.00000	2	0.00000

Table 4. *XPowder©* output quantitative analysis in a thermodiffraction analyses of hidrotalcite. Sum fitted 100 %. The percentages have been calculated weighing data. Angles 2θ have been refined. Normalization criteria: Max. counts = 1.

The possibility to break a diffractogram down into its component parts (figure 4) permits consideration of additional aspects in the interpretation of results. So, some ideas, such as the profile studies (for example Williamson-Hall, 1953 or Warren-Averbach, 1950) require availability of diffractograms of pure phases. In the example of figure 5, difference diagram mainly correspond to dolomitic phase diagram.

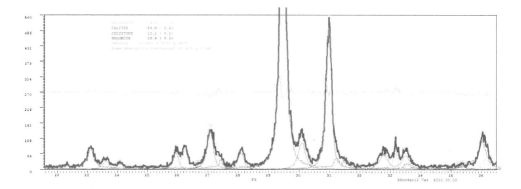

Fig. 4. Detail of the analysis represented in figure 1 that also includes the partially weighted components obtained in the least-squares calculation.

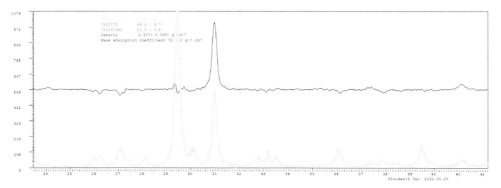

Fig. 5. Isolation of one of the components from the whole diffractogram. In the example the graph of differences coincides with the isolated diffractogram of dolomite, given that this was not included in the calculation and so was the main cause of the differences between the observed and calculated graphics.

3. Adjustement of diffractogram to bar-chart database content

This method can be used when pure phases of sample components are not available, It consists in adjusting the theoretical mixture of the bar-charts found in some databases (PDF2, AMCSD, etc.) to the sample diffractogram. The proportion of each component used in this mixture forms the basis for the quantitative calculation. This method is also contained in the *XPowder©* program, which uses non-linear least-squares methods for the calculation and definition of R_{int} of [6]. The 2θ database content and experimental diagram values can be redefined (see figure 6, table 5).

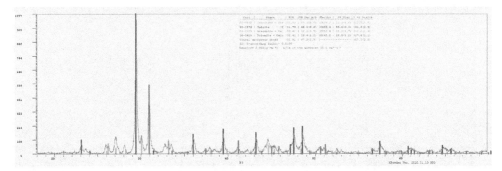

Fig. 6. Graphic *XPowder©* output of quantitative analysis using database patterns.

Given that database records usually show relative intensities (normally up to 100 or 1000), calculations for a given sample must be normalized by means of a simple procedure. After obtaining the diffractogram, we must make a prior calculation of X_P with initial M_P factors = 1. The definitive M_P values of each phase P are then calculated on the basis of the initial X_P values obtained in this first cycle:

Card	Phase	M_P	%W Unc $(\mu/\rho)_P$	σ	$(\mu/\rho)_P$	%W Xtal	σ	%W Xtal+A	σ
01-0885	Celestine	1.00	12.4	1.6	79.2	13.3	1.6	12.3	1.5
83-0578	Calcite	1.70	56.1	0.5	53.4	55.6	0.5	51.5	0.5
01-0628	Aragonite	0.41	12.1	1.7	53.4	12.1	1.7	11.2	1.6
36-0426	Dolomite	2.41	19.4	1.2	45.2	19.0	1.2	17.6	1.1
Global amorphous		2.41	7.9	2.7				7.3	2.2

Table 5. *XPowder©* output quantitative analysis. These data are the inset of the figure 6. Sample F3: R-according factor= 0.0008 Density = 2.940(g cm-3) μ/Dx of the mixture = 55.0 cm² g-1.

$$M_P = (X_P / X_{Reference}) \cdot (W_{Reference} / W_P) \qquad (9)$$

wich is similar to [8]. $W_{Reference}$ is the weight of the reference phase in the standard sample. Likewise, W_P corresponds to the weights of each phase present in the standard sample.

In this fashion, by using a single sample of known composition, we can obtain M_P factors that can be used to quantify any other sample of similar composition. Note that the method can be applied to samples even though not all their components are present.

Normalization process [9] can also be used in the point 2.3, when the samples are recorded in variable experimental conditions.

The program uses database records for least-squares adjustment of the experimental pattern and full-profile RIR factors in order to obtain the results in an absolute scale. The 2θ of both experimental and database patterns were also refined.

3.1 Q-PXRD study applied to atmospheric dust

This type of analysis can also be used to determine the temporal evolution of the quantitative composition of samples taken periodically in any experiment. Because mineral dust is the major contributor to aerosol loading it is interesting to control its temporal evolution. In this

way it has been possible to study atmospheric dust and its temporal evolution (figure 7 and table 6) in the south of the Iberian peninsula as is shown in the following example.

The standard patterns for individual phases were obtained to pure minerals belonging mainly from the Mineralogy Museum, University of Granada, forming thus a database of pure standard phases with initial 'Scale Factors' = 1. The RIR values for the complete profiles were calculated using [9]. The know composition of same samples was used as a standard. In addition, quartz was a common standard for all samples. The goodness of the fit was assessed using the according factor (R_{Int}) for quantification obtained by the XPowder© program using the expression [7], with n=p=1.

$R_{Int} > 0.1$ imply represent good analyses. On the contrary, R_{Int} imply rejection of the analysis and the factors involved in this error must be studied (generally due to experimental causes). In the case of persistent error, the pattern must be rejected.

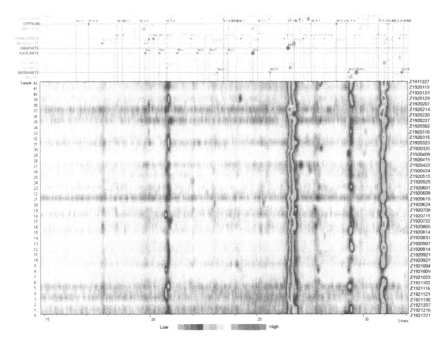

Fig. 7. XPowder© false colour 2D representation of the temporal evolution of mineral phases contained on atmospheric dust deposited on south-eastern Spain along 1992. Note the continuity of quartz during the sampling period.

In this example the following mineral phases have been detected (table 6): bassanite, calcite, chlorite, dolomite, feldspars, graphite, gypsum, halite, kaolinite, illite/muscovite, paragonite, quartz and various smectite types. In addition, some non-classifiable amorphous materials were quantified. From these results interesting conclusions can be draw about the behaviour of mineral aerosols in the atmosphere and enviroment (Díaz-Hernández et al., 2011), as for example the role played by smectites in the atmospheric processing of SO_2 and in secondary sulphate genesis.

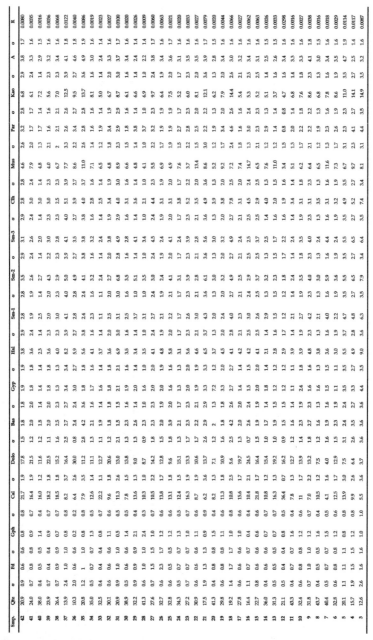

Samp.	Qtz	σ	Fd	σ	Gph	σ	Cal	σ	Dolo	σ	Bas	σ	Gyp	σ	Hal	σ	Sm-1	σ	Sm-2	σ	Sm-3	σ	Clh	σ	Mus	σ	Par	σ	Kao	σ	A	σ	R
42	20.9	0.9	0.6	0.8	0.8	0.8	21.7	0.8	17.8	1.9	1.8	1.5	1.9	1.9	3.8	2.9	2.8	2.8	3.5	2.9	3.1	2.9	2.8	2.8	4.6	2.6	3.2	2.8	6.8	2.9	3.8	1.7	0.0080
41	24.0	0.7	0.7	0.8	0.9	0.7	16.4	0.7	21.5	1.9	2.0	1.2	1.8	1.8	3.6	1.8	1.9	2.5	2.6	1.9	2.6	1.4	3.0	1.7	7.9	2.0	1.7	1.7	6.1	2.4	3.3	1.6	0.0035
40	38.0	0.4	0.5	1.4	1.4	0.4	16.0	0.4	11.6	1.2	1.8	1.2	1.4	1.4	2.3	1.4	2.5	1.4	2.7	1.4	2.0	2.0	3.0	1.4	4.8	1.3	1.7	1.6	7.2	1.4	2.9	1.5	0.0016
39	23.9	0.7	0.4	0.9	0.9	0.7	18.2	0.7	22.5	1.8	2.0	1.1	1.8	1.3	3.6	2.3	2.0	2.0	4.3	2.3	3.0	2.3	3.0	2.3	4.0	2.0	1.6	1.6	5.6	2.3	3.2	1.6	0.0036
38	26.4	0.7	0.7	0.8	0.8	0.7	18.5	0.7	15.2	1.5	2.5	1.6	1.3	1.3	4.0	1.3	2.3	2.3	2.9	2.3	2.8	2.3	3.0	2.3	6.7	2.3	2.1	2.6	7.0	2.3	3.4	1.6	0.0068
37	13.9	2.4	1.0	0.8	0.8	0.8	8.2	0.8	16.4	3.7	2.7	2.5	3.4	2.7	8.2	2.7	4.1	3.9	5.0	3.9	4.1	4.0	5.1	3.9	7.7	3.9	3.3	2.7	12.5	3.7	4.1	1.8	0.0022
36	10.3	2.0	1.0	0.2	0.2	0.8	6.4	0.8	30.0	2.6	2.4	0.8	3.0	1.8	5.9	1.8	2.8	2.8	4.9	2.8	3.5	2.7	3.9	2.7	8.6	3.8	3.4	2.7	7.0	3.6	4.6	1.8	0.0040
35	20.8	1.2	0.6	1.0	1.0	0.5	6.4	0.5	11.2	3.5	4.2	2.5	3.0	1.7	5.6	1.8	2.7	2.4	4.9	3.8	3.8	3.8	4.0	3.7	11.0	2.8	2.8	1.6	13.7	1.4	4.9	1.9	0.0086
33	35.0	0.5	0.7	1.3	1.3	1.0	12.6	0.7	11.1	1.1	2.1	0.8	1.7	1.4	4.1	1.8	3.6	1.6	4.1	1.6	3.2	1.6	2.8	1.6	7.1	1.4	2.8	1.6	8.1	1.4	3.0	1.6	0.0019
32	32.5	0.4	0.4	0.8	0.8	0.8	22.2	0.5	12.7	1.1	1.9	1.3	1.8	1.4	3.7	1.4	2.1	2.0	3.2	2.0	2.4	1.9	3.4	1.4	5.0	1.2	1.9	1.4	6.7	2.0	2.4	1.6	0.0023
31	30.1	0.6	0.6	1.1	1.1	0.6	9.6	0.6	20.6	1.5	1.5	1.2	1.8	2.1	3.6	1.8	2.5	3.0	2.7	2.6	3.8	2.9	3.4	1.9	4.8	1.8	3.4	2.9	5.0	2.0	3.3	1.7	0.0027
30	20.9	0.9	0.6	0.5	0.5	0.5	11.3	0.5	13.0	2.6	2.3	2.1	2.0	1.8	6.9	1.6	3.1	2.5	6.8	3.0	4.9	2.9	4.0	3.0	8.9	2.3	1.8	1.6	8.7	3.0	3.7	1.7	0.0100
29	38.9	0.5	0.9	1.4	1.4	0.6	7.8	0.4	7.8	1.5	2.6	1.6	2.0	2.0	3.5	1.3	3.7	1.4	6.6	1.6	2.8	1.4	4.0	1.6	6.6	1.6	3.8	2.4	6.1	1.4	3.4	1.7	0.0020
28	32.2	0.6	0.5	1.4	1.4	0.4	13.6	0.6	9.0	1.0	2.6	1.4	2.0	1.6	3.4	1.0	2.7	1.0	5.1	1.0	4.1	1.4	3.6	1.4	4.8	1.4	1.4	1.0	6.6	1.4	2.2	1.4	0.0026
27	41.3	0.3	0.6	2.1	2.1	0.7	10.3	0.9	8.7	1.3	2.3	1.1	2.0	2.0	3.5	2.0	3.0	2.4	3.5	2.4	3.4	2.4	3.6	2.3	4.1	1.0	1.7	1.0	6.9	2.4	3.8	1.7	0.0009
26	27.6	0.7	0.7	2.4	2.4	0.6	10.5	0.7	14.2	3.6	2.3	1.8	2.0	1.9	4.1	2.0	4.3	1.9	4.5	2.9	4.5	1.9	4.4	2.3	8.6	2.2	2.3	1.9	6.4	1.9	3.7	1.7	0.0060
25	32.7	0.6	0.6	1.0	1.0	0.6	13.8	0.6	12.8	1.3	2.0	1.5	2.0	2.0	4.8	1.6	2.0	2.4	4.1	2.0	2.4	2.0	3.1	1.9	5.1	1.3	1.6	1.7	7.5	1.8	3.4	1.6	0.0063
24	34.3	0.5	0.5	1.2	1.2	0.5	13.1	0.4	9.6	1.8	3.9	1.3	1.3	1.3	5.8	1.3	2.4	2.1	2.7	1.7	4.1	1.7	3.8	1.7	9.2	1.4	2.0	1.5	5.2	2.0	3.1	1.6	0.0003
23	27.2	0.7	0.7	1.1	1.1	0.6	12.4	0.6	15.1	1.5	2.1	2.0	2.0	1.9	3.1	1.9	2.3	1.7	3.7	2.3	3.9	2.1	5.2	1.7	7.6	1.5	2.9	2.4	6.0	1.7	3.5	1.6	0.0027
22	30.9	0.6	0.4	1.0	1.0	0.6	10.8	0.4	13.3	1.9	2.6	1.7	2.0	1.9	5.6	1.9	3.0	2.1	2.5	1.3	2.5	1.4	4.0	2.0	3.7	2.2	1.3	2.3	6.0	2.9	3.5	1.7	0.0079
21	17.5	1.9	1.3	0.9	0.9	0.6	16.3	0.7	10.6	3.6	2.2	2.9	2.0	3.3	4.6	3.3	4.3	3.0	2.3	1.5	2.8	3.6	2.0	1.3	8.6	2.3	2.2	1.4	6.2	3.3	3.9	1.7	0.0020
20	41.3	0.6	0.4	1.5	1.5	0.8	6.2	0.9	13.7	1.8	2.3	1.4	3.3	2.0	3.7	2.0	2.0	1.2	1.8	1.2	3.0	1.4	3.9	1.4	8.6	0.8	1.3	0.8	7.9	1.4	3.4	1.5	0.0020
19	29.8	0.4	0.6	1.2	1.2	0.6	8.1	0.4	7.1	1.3	2.9	1.6	7.2	1.9	4.5	1.6	2.4	1.4	3.0	1.4	3.2	1.9	4.9	1.4	6.2	1.8	3.4	2.6	14.4	1.8	3.0	1.6	0.0044
18	19.2	1.4	1.4	1.7	1.7	0.8	11.3	0.6	10.9	1.8	2.1	2.5	3.2	2.0	3.7	2.7	4.0	1.9	4.9	2.3	4.9	2.0	3.8	2.0	7.4	1.3	1.6	2.6	5.4	1.4	3.0	1.6	0.0066
17	27.8	1.1	0.6	1.0	1.0	0.4	10.8	0.8	5.6	1.8	4.2	1.3	2.7	2.1	4.2	2.0	2.3	2.0	2.9	2.7	2.4	2.7	7.8	2.5	7.2	1.7	3.0	2.4	5.3	1.8	3.2	1.6	0.0027
16	16.4	0.8	0.6	0.4	0.4	0.6	15.6	0.4	19.7	1.8	2.0	0.7	1.4	1.5	4.2	1.5	3.0	2.1	2.9	2.1	2.1	2.5	4.5	2.4	14.7	2.2	2.2	2.3	5.2	2.3	3.4	1.6	0.0062
15	22.7	0.8	0.6	1.1	1.1	0.6	10.4	0.6	24.5	1.9	2.4	1.5	2.0	2.0	4.1	1.5	2.6	1.5	3.7	2.4	2.5	1.4	2.9	2.5	6.3	1.3	2.3	1.3	5.2	1.4	3.1	1.6	0.0063
14	36.0	0.4	0.5	0.6	0.6	0.7	21.8	0.7	16.4	2.1	2.0	1.5	1.5	2.0	2.1	1.3	2.8	1.3	3.2	2.5	1.4	1.6	4.0	2.0	5.1	1.3	1.3	1.3	6.0	1.4	2.5	1.5	0.0026
13	31.3	0.5	0.5	0.7	0.7	0.7	10.8	0.5	15.4	1.9	1.7	0.9	1.2	1.4	2.9	1.8	1.5	1.2	2.3	1.3	1.7	1.4	2.0	1.6	11.0	1.4	1.4	1.4	3.7	1.6	2.5	1.5	0.0033
12	21.1	0.4	0.4	0.8	0.8	0.7	16.3	0.6	19.2	3.6	1.5	0.9	1.8	1.2	3.9	3.3	2.1	1.2	1.8	1.2	2.2	1.6	3.4	1.6	3.4	0.8	0.8	1.4	4.7	1.4	3.9	1.6	0.0298
11	43.3	0.4	0.7	0.8	0.8	0.6	6.2	0.4	12.7	1.3	1.6	1.4	1.1	1.1	4.8	2.0	2.7	1.4	2.4	1.4	3.5	1.9	3.1	1.8	4.7	1.4	1.8	1.4	6.8	1.8	3.3	1.6	0.0016
10	32.4	0.6	0.8	1.2	1.2	0.6	7.8	0.6	13.9	1.8	2.3	1.6	2.4	1.6	3.9	2.3	2.1	1.9	3.5	2.3	2.4	1.9	3.3	1.6	7.6	2.2	2.0	2.2	7.6	1.8	4.1	1.8	0.0027
9	31.8	0.4	0.5	1.6	1.6	0.4	11	0.4	13.2	1.3	2.2	1.8	1.6	1.5	3.8	1.3	2.7	1.3	3.0	1.3	2.4	1.3	3.5	1.6	6.4	2.2	2.2	1.3	8.6	1.3	3.3	1.5	0.0038
8	43.7	0.4	0.5	1.5	1.5	0.5	7.0	0.6	7.5	1.6	1.9	1.5	1.5	1.1	3.6	1.5	4.0	1.6	5.9	1.9	4.4	1.9	3.2	1.9	6.5	1.7	2.3	1.6	6.8	1.9	3.0	1.5	0.0016
7	40.4	0.5	0.4	1.6	1.6	0.6	4.0	0.6	4.0	1.2	2.3	2.4	3.5	2.7	3.0	1.1	2.1	2.3	3.6	1.9	5.5	1.9	4.9	3.5	11.6	3.1	2.6	1.5	7.8	1.3	3.5	1.6	0.0018
6	32.8	0.6	0.8	1.2	1.2	0.6	12.9	0.6	12.9	1.8	2.4	2.7	3.5	3.2	5.5	2.7	4.7	3.7	5.5	3.6	5.5	3.5	3.2	3.5	6.7	2.3	2.3	1.6	8.6	1.9	4.7	1.9	0.0029
5	20.1	1.1	1.1	1.5	1.5	0.8	9.9	0.8	7.5	3.0	2.7	4.8	3.5	4.9	4.9	2.7	4.8	4.7	5.5	6.5	6.5	5.2	5.2	9.7	2.7	4.1	3.5	11.0	3.5	4.7	1.9	0.0114
4	15.7	1.9	1.4	1.2	1.2	0.6	6.4	0.6	4.7	1.9	2.7	2.7	3.5	3.7	9.0	3.6	6.3	2.7	7.9	6.4	6.4	7.4	4.4	5.2	9.7	2.7	4.4	3.5	14.1	1.6	2.5	1.4	0.0117
3	12.6	2.6	1.6	1.0	1.0	1.0	5.5	1.0	3.7	3.6	3.6	3.6	4.4	3.5	9.0	3.5	6.3	3.4	7.9	6.4	6.4	3.4	7.4	8.1	4.4	3.1	3.1	14.9	3.5	3.2	1.6	0.0097

Table 6. *XPowder*© output quantitative analysis of samples of figure 7. Qtz: Quartz; Fd: Feldespar; Gph: Graphite; Cal: Calcite; Dolo: Dolomite; Bas: Bassanite; Gyp: Gypsum; Hal: Halite; Sm: Smectite; Clh: Clinochlore; Mus: Muscovite; Par: Paragonite; Kao: Kaolinite; A: Amorphous and R: according factor.

3.2 Q-PXRD study applied to potassic volcanic rocks

Volcanic rocks are excellent samples to apply Q-PXRD study because in many cases mineral grain size is very fine and it is difficult establish the modal proportions of minerals that compose the rock, even if study is made using petrographic microscopy. Generally, potassic volcanic rocks have a fine grain size related to fast cooling, among others factors (Mitchell and Bergman, 1991). In these rocks it is necessary know the minerals that form the rock as well as the modal proportions, which are one criteria to classify the sample in the potassic volcanic rocks clan (Mitchell and Bergman, 1991; Le Maitre et al., 2002).

Here we present the result of a Q-PXRDv study of the Zeneta potassic volcanic rocks of the southeast of Spain volcanic region (Fig. 8), classified by many authors as lamproites (Venturelli et al., 1984; Mitchell and Bergman, 1991; Toscani et al., 1995; Duggen et al., 2005; Coticelli et al., 2009). However, recent studies show significant greats mineralogical and geochemical variations, which indicate that the Zeneta volcanic rocks do not correspond to a classical definition of lamproites (Cambeses, 2011).

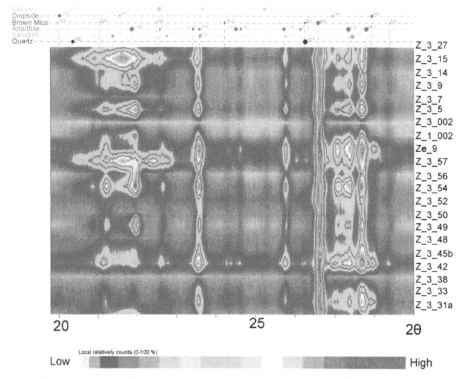

Fig. 8. XPowder© false colour 2D representation of principal mineral phases showing in the most representative samples from rocks of Zeneta volcanic rocks.

Minerals identified in the Zeneta volcanic rocks in the XRD pattern and in petrographic thin section were olivine, diopside, biotite-phlogopite (brown mica), anorthite, sanidine and secondary quartz (Fig. 8). Quantification methodology was performed in fifty-six samples.

Table 7 includes representative analyses. The methodology is based on conversion to absolute values using scale factors for normalization as determined from the diffractogram of a standard with a know composition and modal proportion (sample Z_1_002 in Table 7). The mineral reference was sanidine, for the Zeneta volcanic rocks, due to its high crystallochemical stability its presence in all analyzed samples. The scale factor reference intensity ratios (RIR) were obtained from expression [9]. The goodness of the fit was assesed using the according factor (R_{Int}) for quantification obtained using the *XPowder*© program from the expression [6], using n=p=1.

The results obtained show a very good correlation between Q-PXRD and petrographically determined modal proportions. For this reason we suggest that the XRD results for all samples indicate real petrographic modal proportions (Cambeses, 2011).

Knowing the mineral content of each of the samples permits determination of mineral distribution in the volcanic centre using a simple Krigin interpolation by creating layers with the same isovalues of mineral content based on the quantitative XRD results for each sample.

In Zeneta samples were selected distributed throughout the outcrop (Fig. 9A) and maps were made of all mineral phases identified in the fifty-six quantitative XRD analysis. In present work we show mapping of brown mica, diopside and sanidine distribution (Fig. 9B, C and D).

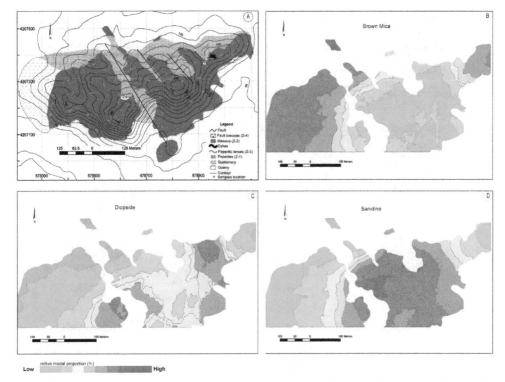

Fig. 9. A) Mapping from Zeneta volcanic rocks outcrop (Cambeses and Scarrow, submitted) B), C) and D) Q-PXRD data interpolated show Brown mica, diopside and sanidine relative modal proportion respectively.

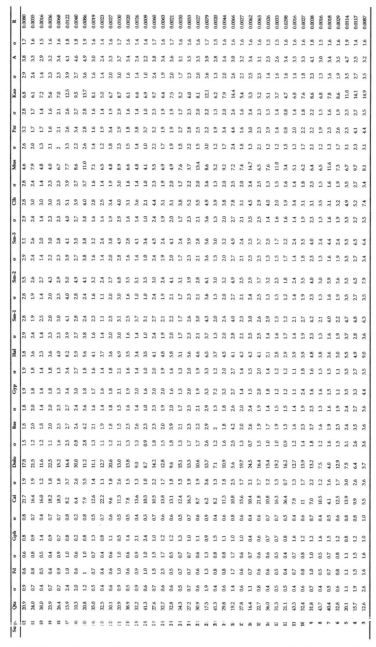

Table 7. Q-PXRD result from Zeneta volcanic rocks of olivine, Ol, diopside, Did, Phlogopite-biotite, Brw-M, Anorthite, An, Sanidine, San, Quartz, Qtz and amorphous, Amor. With σ, as a standard deviation results of adjustment of each mineral. R values is according factor related to adjustment in function of standard diffractogram. Coordinates in UTM projection.

Q-PXRD analyses indicate a heterogeneous distribution of minerals in the Zeneta volcanic rocks. Data obtained reveal how minerals such as diopside and sanidine, typical of lamproites (Mitchell and Bergman, 1991), have their highest modal proportions in central part of the outcrop (Fig. 9C and D). On other hand, brown micas have a higher modal proportion at the edge of the outcrop. Thus a process causing the mineralogical variation may be related to a petrogenetic model for the formation and emplacement of the potassic volcanic rocks (Cambeses, 2011).

For these reasons Q-PXRD analyses may be considered an excellent methodology for the study of volcanic and igneous rocks generally because it is a quick and simple method that provides valuable information about the modal proportions of all the mineral phases present in a sample. Results obtained are in good agreement with conventional methodologies such as modal proportion estimates from petrographic studies. In addition, Q-PXRD results indicate variations in mineral distributions throughout the outcrop potentially providing key information that can be used in petrogenetic interpretations.

4. Conclusions

The use of powerful computers with appropriate software facilitates X-Ray powder diffraction quantitative analysis. Nevertheless, the scientific literature only shows the use of massive semiquantitative analysis in which real measurements of reflective powers are not available because, in part, of the lack of natural standards that are pure enough and have a comparable crystallinity to the samples being studies. In the cases of precise analyses, obtaining good data requires an excessive amount of time in, for example, calculations. In this sense the Rietveld method was a great advance in controlling the experimental parameters and compositions and details of each mineral in mixes. Despite this, the productive capacity of this method is limited and requires a lot of data processing and data interpretation skills. In addition, even with excellent results complications in data processing may obscure data quality.

The least-squares method, applied to the calculation of a diffractogram that is to be compared with experimental data can be simplified as suggested by Rietveld if it is used to calculate experimental diffractograms from a database of a laboratory that applies these methods. This is what was done using methods 2.3 and 3 of the present chapter. Moreover, if pure standards are not available a complex known composition standard may be used to calculate global reference factors (full profile RIR) that can be applied to samples with a similar qualitative composition. Given that programs are available to do these calculations automatically it is possible to do these analyses in a routine manner on natural and industrial substances (online control in industry or mining). Table 8 shows the analytical results for sample F3, this allows the four methods to be compared.

The easiest to use method is 'Simple Quantitative' (semi-quantitative analysis), but this includes large errors although the experimental work and calculations are simple. The Rietveld method gives good results with small standard deviations but a considerable amount of time and effort is required to obtain the results (analytical time 11.7 hours). This necessitates selection of a preferential orientation model and refinement of paramenters which could affect the calculation of the global scale factor for phases with few reflections. The method is not applicable when the structure of some of the sample components is

unknown. In these cases, some programs permit substitution of the structural data for experimental bar graphs if a pure phase is available (e.g., Quanto).

Min\Meth	Simple Q.	Rietveld	σ	Experimental P.	σ	Database	σ
Aragonite (14.5)	12	12.9	0.3	14.3	0.2	12.1	1.7
Calcite (54.1)	49	53.9	0.3	53.8	0.2	55.6	0.5
Celestine (12.6)	8	12.2	0.2	13.2	0.2	13.3	1.6
Dolomite (18.8)	31	21.0	0.2	18.6	0.2	19.0	1.2

Table 8. Comparison of results obtained by different methods of quantification. In brackets % known weight of the mix.

The analyses obtained from the combination of experimental diffractograms show good results and small 'typical deviations', but they are determined by standard analyses (analytical time 12 minutes). It is not necessary to improve the preferential orientation model when standards similar to the analyzed samples are used, so it is possible to obtain reliable results quickly although the method may not be applicable when certain pure phases are missing from the experimental diffractogram. In this case theoretical diffractograms may be used, calculated from crystalline structures and applied to the profile parameters for each phase in the experimental diffractogram of the analyzed sample (distribution function, width of reflections and asymmetry).

The results based on bar models from databases show good results but high deviations because of the variation in experimental conditions with respect to the diffractogram of sample F3. This method can be used routinely and the results are usually sufficiently precise for the majority of industrial processes especially when a bar model is used that was developed in the actual laboratory from, not necessarily pure, samples. This is certainly, in all aspects, the most versatile and balanced method, both experimentally and in terms of data processing.

It is worth underlining about these methods that, with the exception of the first, once the analyses and standard calculations have been performed, from the structural data obtained from experimental results or databases, the analyses of other samples can be performed routinely. For example the program XPowder© permits batch analysis of up to 50 samples simultaneously. As an example, the parameterization of the analyses of atmospheric dust was undertaken in a month. During this time quantitive analysis of various samples was done by image analysis (optical, SEM and TEM) as was chemical analysis, etc. This permitted parameterization of the quantitative analysis and production of the data shown in table 6 in approximately 2 minutes. From now onwards similar samples can be analyzed routinely instantaneously, in real time.

The program used also has numerous others tools for using multiple databases to obtain reliable quantitative analyses, instant refining of unit cells, signal-filtering such as $K_{\alpha2}$ stripping and textural analyses such as Williamson-Hall or Warren-Averbach, etc.

The methods of Q-PXRD applied to the study of igneous rocks permits petrographic characterization of a large number of samples that allows information to be obtained about the petrogenetic evolution of an igneous body.

5. Acknowledgments

This study was supported by the spanish projects CGL2010-16369, CGL2009-09249 and CGL2008-02864 of the Ministerio de Ciencia e Innovación (MICINN), and by the Andalusian grant RNM1595.

6. APPENDIX. The general least-squares methods used by the *XPowder©* program

Consideration on m linear functions (Y_1 to Y_m) with n variables. These variables can be thought of as defining a space whose value at any point is determined by $x_1...x_n$ and the parameters $w_1...w_n$. If the values of the Y functions are measured at n different points (n points for diffraction pattern), we can write:

$$w_1X_{1,1}+w_2X_{1,2}+...+w_nX_{1,n} = Y_1 \text{ (being n>m)} \tag{A1}$$

$$w_2X_{2,1}+w_2X_{2,2}+...+w_nX_{2,n} = Y_2$$

$$w_mX_{m,1}+w_2X_{m,2}+...+w_nX_{m,n} = Y_m$$

The principle of least squares (LS) states that the best values for parameters $w_1...w_n$ are those that minimize the sums of the squares of the differences between the experimental (Yo_i) and calculated (Y_i) values of the function for all the observed (i) points.

Thus, the value to be minimized will be:

$$d = \sum_i \omega_i (Yo_i - Y_i)^2 \text{ with } \omega_i = \text{statistic weight}, \ 1 \leq i \leq m \tag{A2}$$

The minimization problem is treated by differentiating the right-hand side of [A2] in turn and setting the derivative equal to zero:

$$\sum_i w_i (Yo_i - Y_i) \, \partial Y_i / \partial w_k = 0 \qquad \text{with} \qquad 1 \leq k \leq n \tag{A3}$$

Expanding and rearranging:

$$\sum_i \omega_i w_1 X_{i,1} X_{i,1} + \sum_i \omega_i w_2 X_{i,1} X_{i,2} \cdots + \sum_i \omega_i w_m X_{i,1} X_{i,m} = \sum_i \omega_i Yo_i X_{i,1} \text{(A4)}$$

$$\sum_i \omega_i w_1 X_{i,2} X_{i,1} + \sum_i \omega_i w_2 X_{i,2} X_{i,2} \cdots + \sum_i \omega_i w_m X_{i,2} X_{i,m} = \sum_i \omega_i Yo_i X_{i,2}$$

$$\sum_i \omega_i w_1 X_{i,m} X_{i,1} + \sum_i \omega_i w_2 X_{i,m} X_{i,2} \cdots + \sum_i \omega_i w_m X_{i,n} X_{i,n} = \sum_i \omega_i Yo_i X_{i,m}$$

Solution of these m equations gives the best values of w parameters in the least-squares sense.

In a matritial way:

$$X \cdot W = Y \tag{A5}$$

Where the elements of matrix X are defined as:

$$x_{f,c} = \sum_i w_i X_{i,f} \cdot X_{i,c}$$

The elements of matrix $\mathbf{Y}$ are: $y_f = \sum_i w_i Yo_i X_{i,f}$, where

$i = 1, 2,\dots$	n	(pattern points)
$f = 1, 2,\dots$	m	(files)
$c = 1, 2,\dots$	m	(columns)

$\mathbf{X}$ is a square symmetric matrix of know dimensions $(m \cdot m)$, $\mathbf{W}$ is a column matrix of unknowns, of dimension m, and $\mathbf{Y}$ is a know column matrix of dimension m. Multiplying each side of equation [A5] by $\mathbf{X}^{-1}$ gives:

$$X^{-1} \cdot X \cdot W = X^{-1} \cdot Y$$

$$\mathbf{W} = \mathbf{X}^{-1} \cdot \mathbf{Y} \qquad (A6)$$

with the $\mathbf{W}$ column being matrix the solution of a normal equation containing w_i phase weights.

7. References

Alexander, L. & Klug, H.P. (1948). X-Ray diffraction analysis of crystalline dusts. *Analitycal Chemistry*, Vol.20, No.10, pp. 886-894, ISSN 0003-2700.

Bish, D.L. & Howard, S.A. (1988). Quantitative phase analysis using the Rietveld method. *Journal of Applied Crystallography*, Vol.21, No.2, pp. 86-91, ISSN 0021-8898.

Brindley, G.W. (1980). Quantitative X-Ray mineral analysis of clays. In G.W. Brindley and G. Broen, eds. Crystal Structures of Clay Minerals and their X-Ray identification. Mineralogical Society, London, pp. 411-438.

Cambeses, A. & Scarrow, J.H. (submitted). The Mediterranean Tortonian 'salinity crisis', southern Spain: Potassic volcanic centres as potential paleographical indicators. *Geologica Acta*, ISSN-1695-6133.

Cambeses, A. (2011). *Caracterization of the volcanic centres at Zeneta and La Aljorra, Murcia: evidence of minette formation by lamproite-trachyte magma mixing*. Master thesis, University of Granada, 249 pp.

Cardell, C.; Sánchez-Navas, A.; Olmo-Reyes, F.J. & Martín-Ramos, J.D. (2007). Powder X-Ray thermodiffraction study of mirabilite and epsomite dehydration. Effects of direct IR-irradiation on samples. *Analitical Chemistry*, Vol.79, No.12, pp. 4455-62, ISSN 0003-2700.

Chung, F.H. (1974). Quantitative interpretation of X-Ray diffraction patterns of mixtures. I. Matrix-flushing method of quantitative multicomponent analysis. *Journal of Applied Crystallography*, Vol.7, No.6, pp. 519-525, ISSN 0021-8898.

Chung, F.H. (1975). Quantitative interpretation of X-Ray diffraction patterns. III. Simultaneous determination of a set of reference intensities. *Journal of Applied Crystallography*, Vol.8, No.1, pp. 17-19, ISSN 0021-8898.

Clark, G.L. & Reynolds, D.H. (1936). Quantitative analysis of mine dusts: an X-Ray diffraction method. *Industrial Engineering Chemistry, Analytical Edition*, Vol.8, No.1, pp. 36-40, ISSN 0096-4484.

Conticelli, S.; Guarnieri, L.; Farinelli, A.; Mattei, M.; Avanzinelli, R.; Bianchini, G.; Boari, E.; Tommasini, S.; Tiepolo, M.; Prelevic, D. & Venturelli, G. (2009). Trace elements and

Sr-Nd-Pb isotopes of K-rich, shoshonitic, and calc-alkaline magmatism of the Western Mediterranean Region: Genesis of ultrapotassic to calc-alkaline magmatic associations in a post-collisional geodynamic setting. *Lithos*, Vol.107, pp. 68-92, ISSN 0024-4937.

Díaz-Hernández, J.L.; Martín-Ramos, J.D. & López-Galindo, A. (2011). Quantitative analysis of mineral phases in atmospheric dust deposited in the south-eastern Iberian Peninsula. *Atmospheric Environment*, Vol.45, No.18, pp. 3015-3024, ISSN 1352-2310.

Duggen, S.; Hoernle, K.; Van Den Bogaard, P. & Garbe-Schonberg, D. (2005). Post-Collisional Transition from Subduction- to Intraplate-type Magmatism in the Westernmost Mediterranean: Evidence for Continental-Edge Delamination of Subcontinental Lithosphere. *Jounal of Petrology*, Vol.46, No.16, pp. 1155-1201, ISSN 1460-2415.

Le Maitre, R. W.; Streckeisen, A.; Zanettin, B.; Le Bas, M.J.; Bonin, B. & Bateman, P. (Eds.). (2002). *Igneous Rocks. A Classification and Glossary of Terms. Recommendations of the International Union of Geological Sciences Subcommission on the Systematics of Igneous Rocks*. Cambridge University Press, ISBN 0-521-66215-X, Cambridge.

Martín-Ramos, J.D. (2004). Using *XPowder©*, a software package for powder X-Ray diffraction analysis. D.L.GR-1001/04. Spain. ISBN: 84-609-1497-6.

Mitchell, R. & Bergman, S.C. (Eds.). (1991). *Petrology of Lamproites*. Plenum Pres, ISBN 0-306-43556-X, New York

Rietveld, H.M. (1969). A profile refinement method for nuclear and magnetic structures. *Journal of Applied Crystallography*, Vol.2, No.2, pp. 65-71. ISSN 0021-8898.

Toscani, L.; Contini, S. & Ferrarini, M. (1995). Lamproitic rocks from Cabezo Negro de Zeneta: Brown micas as a record of magma mixing. *Mineralogy and Petrology*, Vol.55, No.4, pp. 281-292, ISSN: 1438-1168.

Venturelli, G.; Capedri, S.; Di Battistini, G.; Crawford, A.; Kogarko, L.N. & Celestini, S. (1984). The ultrapotassic rocks from southeastern Spain. Lithos, Vol.17, pp. 37-54, ISSN 0024-4937.

Warren, B.E. & Averbach, B.L. (1950). The effect of cold-work distortion on X-Ray patterns. *Journal of Applied Physics*, Vol.21, No.6, pp. 595-599, ISSN 0021-8979.

Williamson, G.K. & Hall, W.H. (1953). X-Ray line broadening from filed aluminium and wolfram. *Acta Metallurgica*, Vol.1, No.1, pp. 22-31, ISSN 0001-6160.

A Review of Pathological Biomineral Analysis Techniques and Classification Schemes

Maria Luigia Giannossi* and Vito Summa
Laboratory of Environmental and Medical Geology,
IMAA-CNR,
Italy

1. Introduction

The skeleton and teeth are mineralized tissues in the human body. We describe them as physiological biominerals. The origin of them is connected with a precise function in the human body.

In addition to these, there are also biomineralizations occurring spontaneously in the human body, giving rise to disease. These biomineralisations are called pathological biominerals. Urinary stones are an example.

Urinary stones can be composed of an organic matrix mainly containing proteins, lipids, carbohydrates and cellular components, and biominerals.

The compositional analysis of urinary stones is an important requirement for a successful management of the disease, which implies not only a proper evaluation and treatment, but also prophylaxis to prevent recurrence, which is impossible without knowing the composition of the urinary stones involved.

This is a brief review and a comparative study of the principles and practical application of various chemical and physical techniques used for urinary stone analysis.

The different methods for classifying and grouping urinary stones by results of analytic techniques are also compared and evaluated.

2. The pathological biominerals

The term biomineral refers not only to a mineral produced by organisms, but also to the fact that almost all of these mineralized products are composite materials comprised of both mineral and organic components. Furthermore, having formed under controlled conditions, biomineral phases often have properties such as shape, size, crystallinity, isotopic and trace element compositions quite unlike its inorganically formed counterpart.

Biominerals meet the criteria for being true minerals, but they can also possess other characteristics that distinguish them from their inorganically produced counterparts. The

---

* Corresponding Author

most obvious trait is that biogenic minerals have unusual external morphologies. It is perhaps the intricacy and diversity of bio-originated structures that first attracts mineralogically-inclined persons into the field of biomineralization.

A second characteristic of biominerals is that many are actually composites or agglomerations of crystals separated by organic material. In many organisms, they exist as small bodies distributed within a complex framework of macromolecular frameworks such as collagen or chitin (Addadi et al., 2003).

Biominerals can be classified on the same framework as minerals, by composition based on the anionic constituents.

The phenomenon of biomineralisation has been well known for a long time, but insufficiently investigated up to now (Epple, 2002).

The processes of mineralisation in the human body in normal conditions occur in bones and teeth. Bone is the hard connective tissue that forms the skeleton of vertebrates. It consists of layered organized fibrils of collagen impregnated with calcium phosphates containing carbonates and small amounts of other ions, which compose the mineral part of the bone and constitute 60–70% of its weight.

Bones and teeth protect the internal organs, allow enhanced mobility, enable mastication of food, perform other mechanical functions, and are a ready source of the key regulatory inorganic ions calcium, magnesium, and phosphate. The sizes and shapes of bones reflect their function.

The processes of mineralisation in the human body produce also pathological biominerals like arteriosclerosis and kidney or urinary stones.

Formation of crystals in pathological mineralizations follows the same principles as normal calcifications. Local conditions for nucleation require a certain degree of local supersaturation induced by biochemical processes, which can be promoted by deficiency of inhibitors (like diphosphate, magnesium or even citrate ions) and/or the presence of matrix of organic material (such as cholesterol) or other crystals of different solids, that act as heterogeneous nuclei.

While apatite structure minerals are the solid phases found in normal mineralized tissues, pathological calcifications contain several solid phases.

2.1 Urinary stones

Renal lithiasis (renal calculi) affects a wide sector of population, between 4 and 15% approximately (depending on geographical area), and it has been classified as one of the illnesses that can cause much pain to human beings.

Calculi are often heterogeneous, containing mainly oxalate, phosphate, and uric acid crystals. The sequence of events that triggers stone formation is not fully understood yet.

Urinary stones are located in the kidneys, and only a small percentage is lodged in the urinary bladder and urethra.

Kidney stones less than 5 mm in diameter have a high chance of being passed, while those of 5–7 mm have a 50% chance, and those over 7 mm almost always require urological intervention. Renal colic (flank pain) develops as the stone begins its passage down the urinary tract. While approximately 90% of stones are successfully passed out of the urinary tract, the remaining stones generally have to be surgically removed by ureteroscopy or percutaneous nephrolithotomy or comminuted by the non-invasive technique, shock wave lithotripsy.

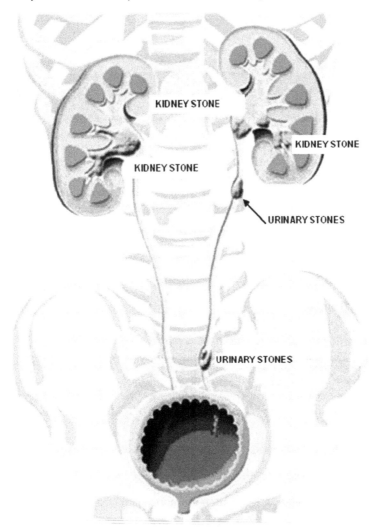

Fig. 1. Urinary and kidney stone localization

General conditions that contribute to stone formation include, e.g., a high concentration of salts in urine, retention of these salts and crystals, pH, infection, and a decrease in the body's natural inhibitors of crystal formation (fig. 2).

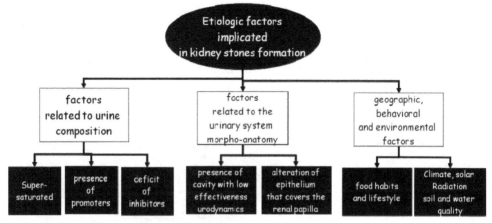

Fig. 2. Etiological factors implicated in kidney stones formation

Many pathways can lead to increased urinary supersaturation. As one example, increased calcium oxalate supersaturation may result from low urine volume or excessive excretion of calcium or oxalate, or combinations of these factors. Hypercalciuria and hyperoxaluria can be result from interaction of genetic susceptibility and environmental triggers, in varying proportions.

Modern medical treatments for stone prevention are largely based on methods to decrease supersaturation effectively, and, thus, doctors are most interested in the pathophysiology leading to specific types of supersaturation.

2.1.1 The minerals inside

The composition of kidney stones can be classified into two parts. The first part is represented by organic matrix containing mainly proteins, lipids, carbohydrates, and cellular components. The other part is biomineral component.

There are some examples of crystalline drug–induced stones that include: indinavir monohydrate, atanazavir sulfate, ceftriaxone (as calcium ceftriaxonate), N4-acetylsulfadiazine, N4-acetylsulfamethoxazole, amoxicillin trihydrate, and triamterene (Schubert, 2000; Dao & Daudon, 1997).

Quartz, calcite, gypsum, and seedcorns are found as artefacts or falsifications among others.

A majority of kidney stones are calcium stones, with calcium oxalate (CaOx) and calcium phosphate (CaP) accounting for approximately 80% of all of these stones, uric acid (UA) about 9%, and struvite (magnesium ammonium phosphate hexahydrate, from infection by bacteria that possess the enzyme urease) approximately 10%, leaving only 1% for all the rest (cystine, drug stones, ammonium acid urate).

Calcium oxalates crystallize has three hydrates— calcium oxalate monohydrate (whewellite), calcium oxalate dihydrate (also known as weddellite) (Sterling 1965; Tazzoli and Domeneghetti 1980), and calcium oxalate trihydrate ($CaC_2O_4 \cdot 3H_2O$; COT) (Deganello et al. 1981), a less common form in pathological stone formation.

Chemical name	Mineral name	Chemical formula	Abbreviation
Oxalates			
• Calcium oxalate monohydrate	Whewellite	$CaC_2O_4 \cdot H_2O$	COM
• Calcium oxalate dihydrate	Weddellite	$CaC_2O_4 \cdot 2H_2O$	COD
Phosphates			
• Calcium phosphate	Hydroxyapatite	$Ca_5(PO_4)_3(OH)$	HA
• Calcium hydrogen phosphate	Brushite	$CaHPO_4 \cdot 2H_2O$	BR
• Magnesium ammonium phosphate hexahydrate	Struvite	$(NH_4)Mg(PO_4) \cdot 6H_2O$	STR
Purines			
• Uric acid		$C5H_4N_4O_3$	UA
• Monosodium urate monohydrate		$N_aC_5H_3N_4O_3 \cdot H_2O$	MSU
Other			
• L-Cystine		$C_6H_{12}N_2O_4S_2$	CY

Table 1. Mineral components of urinary calculi

COM and COD crystals are readily distinguished by their crystal habits (fig. 3): COM usually exhibits hexagonal lozenge morphology, but COD crystallizes as bipyramids, reflecting its tetragonal crystal point group symmetry.

Fig. 3. Calcium oxalate monohydrate and calcium oxalate dihydrate crystals at Scanning electron microscopy view

Sometime stones are made of a single large weddellite crystals (fig. 4), in other cases entire weddellite crystal are transformed into whewellite the most thermodynamically stable form (fig. 5) (Grases et al., 1998).

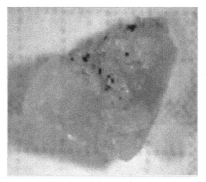

Fig. 4. Weddellite single crystal labelled "spearhead" at stereo-microscopy

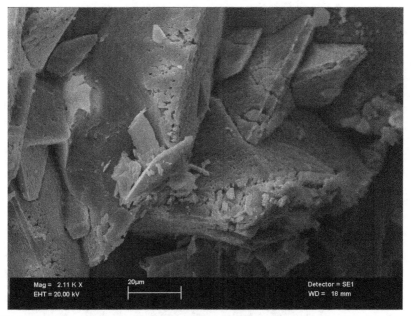

Fig. 5. Weddellite transformation in whewellite, photo at SEM

Various calcium phosphate crystal phases (fig. 6) occur in about one-third of stones, with apatite (apatite is a general term for calcium phosphate in which various anions, e.g. carbonate, fluoride, hydroxide, and chloride, are partially substituted) and brushite found most often admixed with calcium oxalate in an individual stone.

Fig. 6. Phosphates stones at stereo-microscopy observation

Struvite, another mineral entirely constituting some of the stones analyzed, belongs to the group of phosphate so it was difficult to recognize the single struvite crystals at the stereo-microscope, because they appear white and look similar to the other minerals of the same group. Struvite stones are recognizable thanks to their large size and coral form (fig. 7). Despite their porosity, these stones were among the heaviest samples (average weight 155mg).

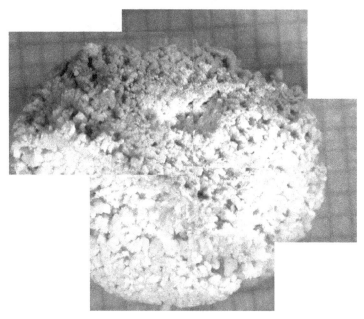

Fig. 7. Struvite stone at stereo-microscopy observation

Cystine stones (fig. 8) are easily recognizable for their appearance: round form and yellow wax colour. The crystals are recognizable by SEM and in thin section by their hexagonal shape. For this type of samples a compact internal structure without porosity was recognized.

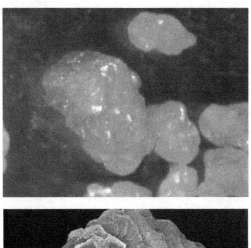

Fig. 8. Cystine at stereo-microscopy and SEM observation

Different types of uric acid crystals (fig. 9) are found in about 10% of stones and are frequently combined with calcium oxalates.

Uric acid monohydrate is very rare, and recently described by the author for the first time (Schubert et al., 2005). The very rare occurrence of a second form of uric acid (Schubert, 1995) could be confirmed by the author. Ammonium urate has a frequency of 1%. The other urate and purine derivates, such as xanthine and dihydroxyadenine, are absolutely rare.

Fig. 9. Uric acid crystals at SEM

3. Techniques for the characterization of kidney stones

Over the first three decades of this century, the stone chemical composition has been investigated only with qualitative methods. These kind of analyses implied wet chemistry qualitative reactions in order to identify the different anions and cations present in the urinary stones. Often this examination is still carried out in the routine clinical laboratory by using specifically designed kits.

More recently, the dissolution of the stones in acidic solution and quantitative measurements of different ions have been performed by atomic absorption spectroscopy or atomic emission spectroscopy with inductively coupled plasma. Many investigators have appreciated the value of quantitative chemical analytic techniques, and several schemes for the classification of urinary stones have been suggested.

On the contrary, in the nineties as for urinary stone analysis there have been a progressive increase in the use of physical techniques (i.e., infrared spectroscopy and X-ray diffraction) and a decrease in the use of chemical methods which by now are regarded as unsatisfactory.

A progressive increase in the use of infrared spectroscopy technique in various biochemical laboratories is observed as it defines the stone composition with accuracy.

This is expected to make procedures easier, revealing more detailed information on mineral structure and the possible identification both of stone type and biomineralisation site.

The X-ray diffraction and crystallographic techniques of polarization microscopy are useful tools in the study of the crystalline structure, order of deposition of components and the nucleus of the urinary stones.

Optical observations can be carried out with a stereomicroscope to determine colour, shape, overall appearance, surface features and any possible occurrence of crystalline layers and/or organic matter on the surface, but the results are only qualitative.

The micro- X-ray analysis which uses an X-ray micro-diffractometer is the most advanced technique. It can be used for classifying urinary calculi by composition as it can detect the multiple stone components and show its structural arrangement, on a whole stone without fragmentation.

	Methods of urinary stone analysis		
	Type of information		
	Chemical composition	Mineralogical composition	Texture
Chemical analysis	full	limited	
Thermal analysis	limited	limited	
SEM	full	limited	full
X-ray diffraction		full	
Infrared spectroscopy		full	
Polarization microscopy		limited	full

Table 2. Methods of urinary stone analysis and their potential information

So far no method has been found to be suitable for providing all the useful information on the structure and composition of urinary stones (tab. 2), only a combination of a refined morphological and structural examination of stones with optical and scanning electron

microscopy, completed with a compositional analysis by using X-ray powder diffraction, can provide a precise and reliable method for the identification of the stone type.

3.1 Chemical methods

The chemical method can identify fairly small amounts of an element but cannot usually identify a compound as such, and in stones of mixed composition, the results merely indicate which ions and radicals are present.

Some of the basic principles (Sutor et al., 1971) of this kind of analysis are the follows:

- A little powdered stone is acidified with 15N hydrochloric acid. Liberation of carbon indicates the presence of carbonate;
- Addition of 20% sodium acetate. A white precipitate indicates the presence of oxalate;
- Addition of ammonium molybdate and 1-amino-2-naphthol-4-sulphonic acid solution. A blue colour shows the presence of phosphate;
- Neutralisation with 5N sodium hydroxide alone. If a white precipitate occurs subsequent addition of 4-nitrobenzene-azo-resorcinol produces a blue colour in the presence of magnesium and a pink colour in the presence of calcium;
- Neutralisation with 5N sodium hydroxide, addition of 15% sodium cyanide and then addition of Folin's uric acid reagent. A blue colour indicates the presence of uric acid;
- Alkalinisation with 5N sodium hydroxide, addition of 15% sodium cyanide and then freshly prepared sodium nitroprusside. A deep purplish colour is obtained in the presence of cystine;
- Neutralisation with 5N sodium hydroxide and addition of Nessler's reagent. A yellow brown colour is formed in the presence of ammonia.

Unfortunately, chemical methods are destructive and need several milligrams of the sample, so small stones cannot be analyzed with chemical methods.

Qualitative and semiquantitative chemical analysis is not accurate and can lead to clinically significant errors (Silva et al., 2010; Westbury, 1989).

Chemical analysis detects calcium and oxalate separately and therefore cannot differentiate crystalline types of CaOx. In a study, COM and COD were evenly distributed (32% each) (Silva et al., 2010). In cystine-containing stones identified by chemical analysis, urate was a major component while cystine was a minor component; however, in the morphological analysis, cystine was a major component. This suggests cystine stones may easily be confused with urate stones if submitted to chemical analysis only.

Currently, chemical analysis of the stones are still practiced but with other methods such as X-ray fluorescence (XRF) spectroscopy and atomic absorption spectroscopy (AAS) or more advanced methods such as SIMS (secondary ion mass spectrometry) (Ghumman et al., 2010).

The most widely practiced chemical analysis, however, are those aimed at identifying not only major elements but minor and trace ones (Trinchieri et al., 2005; Moe, 2006; Atakan et al., 2007; Bazin et al., 2007; Joost & Tessadri, 1987; Meyer & Angino, 1977; Munoz & Valiente, 2005; Sutor, 1969; Welshman & McGeown, 1972). The latter may have played a significant role in urinary stone nucleation and growth, or may be considered as environmental pollution markers (ATSDR, 2008; Bernard, 2008; IPCS, 1992; Jarup, 2002; Patrick, 2003; Satarug et al., 2010).

The major and minor constituents of stones can be investigated by Laser-induced breakdown spectroscopy. The first report, appeared in the literature, on the analysis of biliary stones by LIBS was Singh et al. (2009).

Atomic emission spectroscopy (AES), inductively coupled plasma (ICP), atomic absorption spectroscopy (AAS), neutron activation analysis (NAA), proton-induced X-ray emission (PIXE), and X-ray fluorescence (XRF), require time and labor-intensive specialized sample preparation and presentation protocols for the analysis of elemental composition (Al-Kinani et al., 1984; Zhou et al., 1997).

For fast and in situ analysis, LIBS has been found to be a suitable technique for elemental analysis of any kind of materials (Rai et al., 2002, 2007). The advantage of the LIBS technique is that it does not require any special sample preparation and presentation efforts.

The LIBS technique has proven its own clinical significance for other in vivo applications such as in dental practice for the identification of teeth affected by caries (Samek et al., 2000, 2001). Kumar et al. (2004) have demonstrated LIBS experiments to explore the possibility of using LIBS for in vivo cancer detection.

3.2 Optical and Stereoscopic Microscopy

Binocular stereoscopic microscopy (BSM) is an easily applicable, cost-efficient tool, used to obtain accurate and reliable information regarding the stone components.

Many constituents of renal calculi may be recognized on sight when examining the fractured surfaces under a binocular stereoscopic microscope, permitting a guess as to the probable majoritary composition of the stone.

In practice, the method permits to distinguish between calcium oxalate and calcium phosphate stones. Cystine stones commonly consist of aggregates of well-formed hexagonal prisms or hexagonal tablets and it is very easy to diagnose them with BSM.

BSM was not successful in the analysis of struvite stones.

The analysis of stone composition with microscopic inspection (including polarizing microscopy) is very inaccurate and unfortunately too frequently used for the routine analysis of stones (Herring, 1962; Brien et al., 1982; Prien, 1963). This technique is not capable of identifying small amounts of crystalline materials in admixed samples. A significant contribution to the potentially low level of accuracy using this method is that the accuracy is entirely dependent on the level of sophistication and experiences of the technicians conducting the analyses (Prien, 1963; Silva et al., 2009).

The clinician can perform BSM himself. We believe that any doctor with practical experience can learn to perform an investigation of this kind in a short time. The shape and colour of the stone may provide important information. Following fracturing of the calculus, the order of deposition of components is determined, including identification of an apparent nidus and other patterns, whether homogeneous or characterized by layered, concentric, or radial deposition.

Moreover, what is even more important during BSM analysis is the internal inspection of sections for identifying several structural features, such as the degree of internal organization, the location and size of the nucleus of the stone, the presence of lamination

and/or radial structure in the bulk of the stone, the order of deposition of the components when lamination is present and other structural details.

Likewise, it is possible to distinguish between a sedimentary calcium oxalate monohydrate stone, which shows little or no regularity of the central structure but an outer layer of perfectly developed columnar crystals, and a calculus of the same composition developed by crystal growth which shows a perfectly arranged internal structure.

Some of the best work on the architecture of stones has utilized thin sections of stones, which are studied by optical methods (Murphy & Pyrah, 1962; Cifuentes, 1977). Such studies have elegantly shown the nature of the progressive addition of layers to stones, and have also attempted to identify the nucleus, or initial nidus of the stones (Jung-Sen Liu et al., 2002; Sokol et al., 2003).

The majority of these studies have employed transmission methods of analysis, which require the sample to be present as thin sections approximately 6μm thick (Ouyang et al., 2001; Paschalis et al., 2001; Gadaleta et al., 1996; Mendelsohn et al., 1999, 2000). Unfortunately, thin sections of reproducible thickness are difficult to obtain with urinary stones because of the fragile nature of the material (Murphy & Pyrah, 1962; Cifuentes, 1977).

3.3 Scanning Electron Microscopy (SEM)

Electron microscopy is another method for ultramicroscopic investigation of the fine structure of urinary stones, including single crystal surface structure, sections of urinary calculi, and the possible presence of unknown components within the calculus (Hesse et al., 1981; Hyacinth et al., 1984). However, it also needs specialized equipment. The material in urinary calculi is also prone to irradiation damage during electron microscopy and this suggests the need for care in the interpretation of data (Crawford, 1984).

Scanning electron microscope uses electrons rather than light to form an image (Walther et al., 1995). It has a large depth of field, which allows a large amount of the sample to be focused at a time. SEM produces images of high resolution, which means that closely spaced features can be examined at a high magnification (Harada et al., 1993). Preparation of sample is relatively easy since most SEMs only require the sample to be conductive (Lee et al., 2004). The spatial distribution of major and trace elements can be studied in a range of human kidney and bladder stones with well-documented histories to understand their initiation and formation

3.4 X-ray diffraction analysis

There is no doubt that XRD is the most-appropriate method to determine mineral structures. XRD can distinguish all the different crystal types in a particular mixture, and is therefore accepted as the gold standard for stone analysis (Ghosh et al., 2009; Giannossi et al., 2010). However, this method is not easily accessible. Furthermore, it is expensive, requiring specialized equipment and trained staff.

XRD has advantages of reliability in qualitative analysis and accuracy in quantitative analysis. It operates simply and has a high sensitivity. Based on XRD diffraction data, the multicomponents in a sample can be measured simultaneously.

Another advantage of the X-ray powder diffraction technique is that the powder can be characterized without a surgical procedure by analyzing the fragmented crystals collected from the urine, which follows the extra-corporeal shock wave lithotripsy ECSWL.

Virtually all crystal structures are unique is some structural aspect and their diffraction patterns can be differentiated from other structures and diffraction patterns (fig. 10). Highly sensitive and accurate XRD instruments are often necessary to differentiate some of the structures seen in stones as their chemistry and crystal structures can be similar.

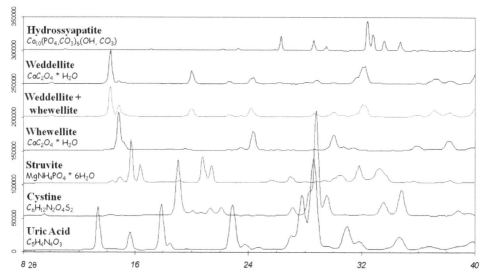

Fig. 10. XRD profiles

The XRD data for the most common components of human kidney stones are presented in Table 3.

COM		COD		AP		ST		UA		CY	
d (Å)	I	d (Å)	I	d (Å)	I	d (Å)	I	d (Å)	I	d (Å)	I
5.93	100	8.70	12	8.15	29	5.90	38	6.56	38	4.70	100
5.79	25	6.31	100	5.26	6	5.60	45	5.63	20	4.68	100
4.64	7	6.15	45	4.08	8	5.38	22	4.91	48	4.63	100
4.52	6	4.40	45	3.89	10	4.60	6	4.76	6	4.45	5
3.78	13	3.89	14	3.44	45	4.25	100	3.86	55	4.20	7
3.65	100	3.09	18	3.17	10	4.14	34	3.70	5	3.32	8
3.00	10	3.07	18	2.81	100	3.29	24	3.28	14	3.18	19
2.97	46	2.81	20	2.78	62	3.02	10	3.18	70	3.13	30
2.91	12	2.77	20	2.72	58	2.96	16	3.09	100	3.06	16
2.89	10	2.75	85	2.63	31	2.92	46	2.87	25	2.71	16
2.84	14	2.41	14	2.53	5	2.80	26	2.80	10	2.70	32
2.48	30	2.39	14	2.40	7	2.72	8	2.57	18	2.68	32
2.34	90	2.33	10	2.26	20	2.69	37			2.60	17

Table 3. XRD data for the most common components of human kidney stones

The diffraction data are presented in crystallographic language as interplanar d-spacings in Ångstroms (d(Å)) associated with the distances between atoms in the structure and as diffraction intensities either relative to a weak versus strong scale or on a maximum of 100 scale (Sutor et al., 1968; Mandel & Mandel, 1982; JCPDS, 1985).

In practice, the experimental XRD patterns are compared with those for the standard patterns presented in Table 3. All diffraction lines of a given standard pattern, especially the strongest lines, must be matched with diffraction lines in the sample pattern. If some lines of a given intensity are thought to match a standard, then all lines with equal or greater intensity must also match the standard lines. Remaining unmatched lines are used to determine any other crystalline components in the sample. With the advent of high resolution XRD cameras utilizing focusing monochromators and high flux X-ray generators, the ability to detect minor stone components has greatly increased as the diffraction patterns appear sharper and diffraction lines are easily differentiated from neighbouring diffraction lines separated by as little as 0.01–0.03 Å interplanar spacings.

Renal calculi with calcium oxalates are represented by the general formula CaC_2O_4 xH_2O, where x is the number of bonded-water molecules, which can vary from 1 to 2. It can be formed on crystalline seed particles of organic or inorganic compounds that work as a nucleating substrate. Therefore, the H_2O molecule might be bound or free, depending on if the H_2O molecule belongs to the crystal structure or the organic compound among them. Some of the characterization techniques commonly used are not suitable to give the structural information about the H_2O molecule.

The increased sensitivity has allowed for the identification of smaller amounts of poorly crystalline materials such as apatite. Unfortunately, if the stone material is a drug, or drug metabolite whose XRD pattern or single crystal structure has not been published, XRD methods fail to definitively characterize the sample. In those cases, the XRD method can only tell you what the stone is not composed of.

Many modern methods of analysis, also powder XRD, destroy the structure of the calculi when the samples are prepared for introduction into the instrument. Maintaining the structural integrity of the calculi is important for the elucidation of the chemistry of formation and the etiology of the calculi in the urinary system.

The micro-diffractometer XRD is preferable used when a very limited amount of sample is available but also on the bulk sample, without any type of treatment.

3.5 Fourier Transform Infrared (FT-IR) spectroscopy

Several reports have been published on the comparison of IR techniques to wet chemical methods for renal stone and other biological analysis, though these can be somewhat outdated (Anderson et al., 2007; Gault et al., 1980; Carmona et al., 1997).

Infrared spectroscopy was first applied to stone analysis by Beischer in the mid fifties (1955).

Weissman et al. (1959), Klein et al. (1960), Tsay (1961), Takasaki (1971), and Modlin (1981) have performed analysis of renal stones by IR with paste and KBr table method. Bellanato et al (1973) have identified with IR the different types of oxalates, phosphates and urate in urinary stones. Oliver and Sweet (1976), proposed a systematic scheme for the qualitative identification and interpretation of the IR spectra which was applied by Gault et al. (1980),

and compared with wet chemical analysis. It is a useful technique for identifying organic and inorganic compounds. In fact, it is particularly useful for determining functional groups present in a molecule, because they vibrate at nearly the same frequencies independently on their molecular environment.

Like X-ray diffraction, infrared spectroscopy provides results on the actual salts, including the different degree of hydratation, with an additional advantage of identifying non crystalline compounds, whereas X-ray diffraction cannot. Moreover, recent advances in computerized infrared spectroscopy, particularly Fourier transform infrared (FT-IR) spectroscopy, have allowed to obtain infrared spectra in less than a minute, whereas in a conventional X-ray apparatus each run requires some hours. Finally, the quantity of sample needed for Fourier transform infrared spectroscopy can be less than one microgram.

In FTIR spectral analysis, spectral data is related to the vibrational motions of atoms in bonds (e.g., bond stretching, bond contracting, or bond wagging, etc.).

Classically, the powdered sample is admixed with powdered potassium bromide, compressed into a nearly transparent wafer, and the IR beam is passed through the wafer. Recently, advances in other sample preparation methods have allowed powdered samples to simply be ground to ensure optimal sampling of a multicomponent stone and then the IR beam is directed at the sample surface (attenuated total reflectance). Although FT-IR can yield qualitative and quantitative results, the preparation of the calculi samples is time consuming and difficult.

The reflected IR beam containing spectral data specific to the sample is then recorded.

The IR pattern contains absorption bands representing specific energies (presented as wavelengths in units of cm^{-1}, or more commonly known as wavenumbers) corresponding to molecular motions in molecules. It is therefore possible to differentiate molecular motions in similar organic groups. The IR pattern of a mixed component stone is frequently very complex, but the advent of computer controlled IR spectrometers, especially modern FTIR spectrometers has allowed for computer assisted pattern stripping and comparative standards library matching.

For XRD and FTIR, the accuracy of the analysis is very strongly dependent on the quality of standard spectra. Most laboratories conducting stone analyses prepare their own standards libraries. Unfortunately, many analysis laboratories use patient stone material to create their standard spectra. As their stones are analyzed by the same method as they are using to analyze other stone samples, their unknowns become their standard. As virtually no stone is composed of only one pure crystalline component, such spectral libraries are very inaccurate and the potential for skewed and inaccurate stone analysis is highly probable. Preparation of synthetic stone components for the generation of standards and verification of composition by alternative methods is the only correct way to prepare a standards library for either XRD or FTIR, especially for FTIR. Commercial libraries should only be used for supplemental data in those rare instances when experimental data cannot be correlated with defined stone component standards, especially for identification of nonbiologic or false stones.

Identification is very simple if a reference spectrum that matches that of the unknown material is found. When an exact reference spectrum match cannot be found, a band by band assignment is necessary to determine the composition of the solid.

Infrared spectroscopy permits to clearly distinguish between a calcium oxalate monohydrate renal calculus and a calcium oxalate dihydrate renal calculus. Thus, absorption bands comprised between 3500 cm^{-1} and 750 cm^{-1} are clearly different for both compounds (Daudon et al., 1993).

All phosphate containing calculi show an intense absorption band around 1000 cm^{-1}. This band permits its easy identification even in mixtures with calcium oxalate monohydrate or dihydrate. Pure brushite calculi are not frequent, but they exhibit characteristic IR spectra that allow to clearly distinguish them from hydroxyapatite or ammonium magnesium phosphate calculi.

Uric acid is probably one of the cases where a wider variety of sizes and colours can be found, and consequently important mistakes can be produced if the identification is exclusively performed visually. The infrared spectra of such calculi are, nevertheless, characteristic and permit their easy identification without any difficulty and also allow their clear differentiation from the infrared spectrum corresponding to ammonium urate calculi due to the different absorption bands comprised between 1300 cm^{-1} and 500 cm^{-1}.

The real benefit of FTIR is the high sensitivity of the new computer controlled spectrometers that can take many repetitive spectra of the same sample and mathematically enhance the sample signal to experimental noise ratio.

3.6 Thermal analysis

The thermal decomposition and structural study of biological materials—urinary calculi (Kaloustian et al., 2002; Afzal et al., 1992; Madhurambal et al., 2009), enamel and dentin (Holager, 1970), and bones (Paulik et la., 1969; Mezahi et al., 2009; Mitsionis et al., 2010)— have been studied many times.

The thermal study of kidney stones has been published (Strates et al., 1969; Ghosh et al., 2009): differential thermal analysis (DTA), thermogravimetry (TG), differential scanning calorimetry (DSC), can, also, characterize the main components (alone or in mixture) in urinary calculi.

When stones are mixtures of the two oxalates hydrates, it is difficult, to differentiate calcium oxalate monohydrate (COM, Whewellite) and calcium oxalate dihydrate (COD, Weddelite) in the binary mixtures, except when one of them is in little quantity in the calculi. A very low heating rate by DSC (0.3°C min–1, from 100 to 180°C) permitted the differentiation of the two hydrate forms.

Under nitrogen sweeping, the TG, DTG (derivative curve of the thermogravimetry) and DTA curves of the COM standard, display three typical steps, located in the temperature ranges of about 100-220, 450-520 and 600-800°C . In the thermal curves of a calcium oxalate dihydrate sample, two endothermic peaks, attributed to the water volatilization, are near 164 and 187°C. Then the same curves (DSC, TG) as for COM were observed (Farner & Mitchell, 1963; Berényi & Liptay, 1971).

A simultaneous thermal analysis apparatus (TG-DTA) was usually used, with: heating rate 5°C min–1, from the ambient temperature to 850-1230°C, gas sweeping: air (0.5 L h^{-1}) or nitrogen (2.5 L h^{-1}). Thermocouples and crucibles were platinum. The sample mass ranged from 3.7 to 10 mg, and kaolin or α-Al_2O_3 (Merck) was used as an inert thermal reference.

The thermal study can, also, characterize the magnesium ammonium phosphate hexahydrate or struvite, and uric acid (UA).

The average of the peak temperatures, computed from urinary stones of struvite, were, 108 and 685°C. These values are very near those of the struvite standard.

The average temperatures from urinary stones of UA, were 418 and 446°C showing difference with the standards presenting higher values: 429 and 450°C.

3.7 Imaging investigations

The micro computed tomography (micro CT) as a potential method for the analysis of urinary stone composition and morphology in a nondestructive manner at very high resolution (Zarse et al., 2004). Micro CT, which has seen considerable use as a research tool in bone biology (Ruegsegger et al., 1996), has the ability to reconstruct 2-D and 3-D images of urinary stones that allow the 3-D image of the stone to be cut and viewed in multiple planes with voxel sizes of 8–34 μm.

Micro CT allows non-destructive mapping of the internal and surface structure of urinary stones and permits identification of mineral composition based on x-ray attenuation values. Micro CT cannot differentiate mineral types when the stone is highly complex and micro-heterogeneous with significant mixing of different mineral types at a scale below the spatial resolution of the instrument.

4. Type of kidney stones: Classification

Despite the many results achieved with all these techniques, very little attention has been paid to the classification scheme to show a clear correlation with pathogenesis, structure and composition of calculi.

Morphological and textural data are very significant and recent classifications also deal with this kind of observations to distinguish eight types of urinary stones and at least 30 sub-categories (tab. 4).

On the contrary, previous categories were distinguished only on chemical bases (oxalate, phosphate, urate and cystine). This tends to underestimate the complexity of an individual's stone history as, indeed, it has been determined that the vast majority of stones actually contain more than one type of mineral.

The papers published in the past (Brien et al., 1982; Elliot, 1973; Herring, 1962; Murphy & Pyrah, 1962; Kim, 1982; Leusmann, 1991) must be considered the first step for making a fundamental tool in clinical uses. Finally, in 1993, Daudon et al. (1993) established the first classification of renal calculi with a clear correlation with the main urinary etiologic conditions. However, this information is complex and probably is difficult to adapt to clinical routine practice, in spite of its interest for scientific purposes. Consequently, it was necessary to establish a classification of renal calculi, in accordance to its composition and fine structure, clearly correlated with specific pathophysiologycal conditions as the main urinary alterations, adapted to the common clinical practice.

The latest classification scheme suggested (Grases et al., 1998, 2002) is very detailed and is useful for classifying each type of kidney stone, and, therefore, each patient in more than 30

different subgroups characterized by specific etiologic factors necessary to determine the treatment and disease prevention, especially in the presence of mixed stones requiring a proper intervention for each mineral phase present.

This classification constituted the first attempt to set up a classification of renal calculi useful for clinical purposes and also the first effort to find the relationships between pathogenesis, structure and composition of calculi, yet no connections with urinary parameters were established.

GROUP	Description	TYPE	Description	SUBTYPE	Description
1	Calcium oxalate monohydrate (whewellite) - papillary kidney stone	1a	core constitued by whewellite / organic matter	1aI	core constitued by organic matter
				1aII	core constitued by weddellite and organic matter
		1b	core constitued by hydrxyapatite/organic mater	1bI	core constitued by hydroxyapatite
				1bII	core constitued by hydroxyapatite and organic
2	Calcium oxalate monohydrate (whewellite) - kidney stone in cavity	2a	core constitued by whewellite + organic matter		
		2b	core constitued by hydroxyapatite + organic matter		
		2c	core constitued by uric acid		
3	Calcium oxalate dihydrate (weddellite)	3a	weddellite only	3aI	without transformation in whewellite
				3aII	with transformation in whewellite
		3b	hydroxyapatite in small quantities	3bI	core constitued by hydroxyapatite
				3bII	containing little amounts of hydroxyapatite among weddellite crystals
				3bIII	containing little amounts of hydroxyapatite and organic matter among weddellite crystals
		3c	papillary		
4	Weddellite + Hydroxyapatite mixed stone			4I	alternative weddellite/hydroxyapatite layers
				4II	disordered weddellite/hydroxyapatite deposits
5	Hydroxyapatite	5a	hydroxyapatite only		
		5b	weddellite in small quantities		
6	Struvite				
7	Brushite				
8	Uric Acid	8a	uric acid only	8aI	compact uric stone
				8aII	layered uric stone
				8aIII	disordered uric stone
		8b	uric acid + uric acid dihydrate	8bI	alternative anhydrous/dihydrate uric acid layers
				8bII	disordered anhydrous/dihydrate uric acid deposits
		8c	urates		
9	Whewellite + uric acid mixed stone			9I	papillary stone
				9II	unattached (no papillary) stone
10	Cystine				
11	Infrequent stones	11a	organic matter as main components		
		11b	medicamnetous		
		11c	artefacts		

Table 4. Classification scheme

5. Stone analyses procedures

Because most stones are multicomponent, the method employed in the analysis of stone material should be capable of resolving all components of the stone, especially all the crystalline components.

The literature on stone analysis methods clearly supports the use of XRD or FTIR as the prime choices. One issue not yet resolved in the literature is the level of accuracy one should accept in analysis reports and the rank order of compositional analysis in multicomponent stones.

A possible example of a recommended procedure to analyse urinary stone is explain in the figure 11.

The mission of the laboratory is to provide information necessary for clinical decision making and patient care.

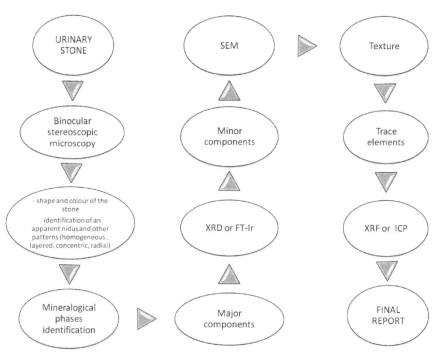

Fig. 11. Flow pattern of urinary stone analysis

Laboratory analyses generate multiple different data types that may include text, quantitative, graphic, and digital image data. Combining the different types of data produced during laboratory analyses into a comprehensive report can maximize the effectiveness of the information presented to clinicians who are relying on the report to guide diagnostic and therapeutic decisions.

Unfortunately, these data types often reside in multiple separate systems, and integrating them into a report often requires laborious procedures, which are inefficient and fraught with potential for error. The management of data produced during kidney stone analysis is an example of such a situation.

The KISS system developed by Shang-Che Lin et al. (2002) is a good example to integrate patient and specimen information from the laboratory information system, digital images of stones, and analytic instrument data into a concise report for the ordering clinicians. The database management environment facilitates archival and retrieval capabilities. Implementation of the system has reduced the number of manual steps necessary to produce a report and has saved approximately 30 technologist hours per week. Transcription errors have been virtually eliminated.

6. Conclusion

Table 5 shows a summary of the comparative assessment of the various methods of stone analysis. It may be inferred that any of these methods is only as good as he sample used, and different areas of the stone must be analyzed separately if useful results are to be obtained.

While the wet chemical analytical qualitative method of urinary stone remains the traditional gold standard, these have been increasingly globally replaced with the more accurate and quantitative methods, such as infrared spectroscopy and X-ray diffraction.

Unfortunately, many urologists make no use of stone analysis due to cost reasons, ignorance, or convenience.

	Chemical analysis	Thermal analysis	SEM	X-ray diffraction	Infrared spectroscopy	Polarization microscopy
Relative cost factor	****	***	*	**	**	****
Analysis Time	**	**	*	***	****	*
Sample preparation	**	**	*	***	***	****
Degree of accuracy	**	***	***	****	***	*

Table 5. Analysis of different methods of urinary stone analysis (****= good;   *=bad)

7. References

Addadi L., Raz S., Weiner S. (2003). *Taking advantage of disorder: amorphous calcium carbonate and its roles in biomineralization.* Adv Mat 15:959-970

Afzal M., Iqbal M., Ahmad H., (1992). Thermal analysis of renal stones. *J Therm Anal.* 38:1671-82

Al-Kinani A.T., Harris I.A., Watt D.E., (1984). Analysis of minor and trace elements in gallstones by induction of characteristic ionizing radiation. *Phys Med Biol* 29:175-184

Anderson J.C., Williams J.C. Jr, Evan A.P., Condon K.W., Sommer A.J. (2007). Analysis of urinary calculi using an infrared microspectroscopic surface reflectance imaging technique. *Urol Res* 35:41-48

Atakan I.H., Kaplan M., Seren G., Aktoz T., Gül H., Inci O., (2007). Serum, urinary and stone zinc, iron, magnesium and copper levels in idiopathic calcium oxalate stone patients. *Int Urol Nephrol* 39:351-356

ATSDR, (2008). *Toxicological Profile for Cadmium (Draft for Public Comment).* US Department of Health and Human Services, Public Health Service, Atlanta.

Bazin D., Chevallier P., Matzen G., Jungers P., Daudon M., (2007). Heavy elements in urinary stones. *Urol Res* 35:179-184

Beischer D.E., (1955). Analysis of renal calculi by infrared spectroscopy. *J Urol;*73:653-659

Bellanato J., Delatte L. C., Hidalgo A., Santos M., (1973). *Application of infrared spectroscopy to the study of renal stones, in Urinary Calculi: Recent Advances in Aetiology, Stone Structure and Treatment:* Proceedings of the International Symposium on Renal Stone Research, edited by L. C. Delatte, A. Rapado, and A. Hodgkinson S. Karger AG, Basel, Switzerland, pp. 237-246

Berényi M., Liptay G., (1971). The use of thermal analysis in medical science with special reference to nephroliths. *J. Therm. Anal.*, 3, pp. 437–443

Bernard A., (2008). Cadmium and its adverse effects on human health. *Indian J. Med. Res.* 128, 557–564

Brien G., Schubert G., Bick C., (1982). 10,000 analyses of urinary calculi using x-ray diffraction and polarizing microscopy. *Eur Urol*; 8: 251–256

Carmona P., Bellanato J., Escolar E., (1997). Infrared and Raman spectroscopy of urinary calculi: A review. *Biospectroscopy* 3:331–346

Cifuentes D.L., (1977). Study of calculi structure using thin mineralogical section. *J Urol Nephrol* 83(2):592–596

Crawford D., (1984). Electron microscopy of urinary calculi – some facts and artefacts. *Urol Res* 12:17–22

Dao N.Q., Daudon M., (1997). *Infrared and Raman Spectra of Calculi*. Paris: Elsevier; 1997.

Daudon M., Bader C.A., Jungers P., (1993). Urinary calculi: review of classification methods and correlations with etiology. *Scanning Microsc*;7:1081– 106

Deganello S., Kampf R.A., Moore B.P. (1981). The crystal structure of calcium oxalate trihydrate: Ca(H2O)3(C2O4). *American Mineralogist*, 66, 859-865

Elliot J.S., (1973). Structure and composition of urinary calculi. *J Urol*;109:82– 3

Epple M., (2002). In: *Proceedings of the Workshop 'Grundlegende Aspekte der Biomineralsation*, Bochum, (Ed.), p. 1. (A contribution to understand...)

Farner V.C., Mitchell B.D., (1963). Soil Sci., 96: 221., in *Differential Thermal Analysis* , R. C. Mackenzie, Academic Press, London, 1970 Vol. 1

Gadaleta S., Landis W., Boskey A., Mendelsohn R., (1996). Polarized FT-ir microscopy of calcified turkey leg tendon. *Connect Tissue Res* 34:203–211

Gault M.H., Ahmed M., Kalra J., Senciall I., Cohen W., Churchill D., (1980). Comparison of infrared and wet chemical analysis of urinary tract calculi. *Clin Chim Acta* 104:349–359

Ghosh S., Basu S., Chakraborty S., Mukherjee A.K.., (2009). Structural and microstructural characterization of human kidney stones from eastern India using IR spectroscopy, scanning electron microscopy, thermal study and X-ray Rietveld analysis. *J. Appl. Cryst.* 42, 629–635

Ghumman C.A.A., Carreira O.M.T., Moutinho A.M.C., Tolstogouzov A., Vassilenko V., Teodoro O.M.N.D., (2010). Identification of human calculi with time-of-flight secondary ion mass spectrometry. *Rapid Commun. Mass Spectrom.*; 24: 185–190

Giannossi M.L., Summa V., (2010). New mixed urinary stone: review of classification scheme. *The Urologist*. (under review)

Grases F., Costa-Bauzá A., Ramis M., Montesinos V. & Conte A., (2002). Simple classification of renal calculi closely related to their micromorphology and etiology. *Clin Chim Acta* 322: 29-36

Grases F., Garcı´a-Ferragut L., Costa-Bauza´ A., (1998). Analytical study of renal calculi. A new insight. *Recent Res Dev Pure Appl Anal Chem*;1:187– 206

Harada Y., Tomita T., Kokubo Y., Daimon H., Ino S., (1993). Development of an ultrahigh vacuum high-resolution scanning transmission electron microscope. *J Electron Microsc (Tokyo)* 42:294–304

Herring L.C., (1962). Observations on the analysis of ten thousand urinary calculi. *J Urol*; 88: 545–562

Hesse A., Hicking W., Bach D., Vahlensieck W., (1981). Characterisation of urinary crystals and thin polished sections of urinary calculi by means of an optical microscopic and scanning electron microscopic arrangement. *Urol Int* 36:281–291

Holager J., (1970). Thermogravimetric examination of enamel and dentin. *J Dent Res*. 49:546–8

Hyacinth P., Rajamohanan K., Marickar F.Y., Koshy P., Krishnamurthy S., (1984). A study of the ultrastructure of urinary calculi by scanning electron microscopy. *Urol Res* 12:227–230

IPCS, (1992). *Environmental health criteria 134, Cadmium*. World Health Organization, Geneva.

Jarup L., (2002). Cadmium overload and toxicity. *Nephrol. Dial. Transplant.* 17 (Suppl. 2), 35–39

JCPDS, (1985). *Powder Diffraction Files*. International Centre for Diffraction Data. Swarthmore, PA, 1985

Joost J., Tessadri R., (1987). Trace element investigations in kidney stone patients. *Eur Urol* 13:264–270

Jung-Sen Liu, Ching-Shui Huang, And Heng-Hui Lien., (2002). Structural analysis of gallstones with thin-section petrographic microscopy: A study of 100 gallstones from Taiwanese patients. *J Lab Clin Med*;140:387-90

Kaloustian J., Pauli A.M., Pieroni G., Portugal H., (2002). The use of thermal analysis in determination of some urinary calculi of calcium oxalate. *J Therm Anal Calorim.*;70:959–73

Kim K.M., (1982). The stones. *Scanning Electron Microsc*;IV: 1635– 60

Klein B., Weissman M., Berkowitz J., (1960). Clinical applications of infrared spectroscopy. II. Identification of pathologic concretions and other substances. *Clin Chem.*, 6: 453-465

Kumar A., Yueh F.Y., Singh J.P., Burgess S., (2004). Characterization of malignant tissue cells by laser- induced breakdown spectroscopy. *Appl Opt* 43:5399–5403

Lee K.M., Cai Z., Griggs J.A., Guiatas L., Lee D.J., Okabe T., (2004). SEM/EDS evaluation of porcelain adherence to gold-coated cast titanium. *J Biomed Mater Res Part B Appl Biomater* 68(2):165–173

Leusmann D.B., (1991). A classification of urinary calculi with respect to their composition and micromorphology. *Scand J Urol*; 25:141–50

Madhurambal G., Subha R., Mojumdar S.C., (2009). Crystallization and thermal characterization of calcium hydrogen phosphate dihydrate crystals. *J Therm Anal Calorim.*;96:73–6

Mandel N.S., Mandel G.S., (1982). Structures of crystals that provoke inflammation. In: *Advances in Inflammation Research*, (Weissmann G, ed.). Raven Press, New York, NY, pp. 73–94

Mendelsohn R., Paschalis E., Boskey A., (1999). Infrared spectroscopy, microscopy, and microscopic imaging of mineralizing tissue: spectra-structure correlations from human iliac crest biopsies. *J Biomed Opt* 4:14–21

Mendelsohn R., Paschalis E., Sherman P., Boskey A., (2000). IR microscopy paging of pathological states and fracture healing of bone. *Appl Spectrosc* 54:1183–1191

Meyer J., Angino E.E., (1977). The role of trace metals in calcium urolithiasis. *Invest Urol* 14:347–350

Mezahi F.Z., Oudadesse H., Harabi A., Lucas-Girot A., Le Gal Y., Chaair H., Cathelineau G., (2009). Dissolution kinetic and structural behaviour of natural hydroxyapatite vs. thermal treatment. *J Therm Anal Calorim.*;95:21–9

Mitsionis A.I., Vaimakis T.C., (2010). A calorimetric study of the temperature effect on calcium phosphate precipitation. *J Therm Anal Calorim.*;99:785–9.

Modlin M., Davies P.J., (1981). The composition of renal stones analysed by infrared spectroscopy. *S. Afr. Med. J.*, 59: 337-341

Moe O.W., (2006). Kidney stones: patophysiology and medical management. *Lancet* 367:333–344

Munoz J.A., Valiente M., (2005). Effects of trace metals on the inhibition of calcium oxalate crystallization. *Urol Res* 33:267–272

Murphy B.T., Pyrah L.N., (1962). The composition, structure, and mechanisms of the formation of urinary calculi. *Brit J Urol*; 34:129– 85.

Oliver L.K., Sweet R.V., (1976). A system of interpretation of infrared spectra of calculi for routine use in the clinical laboratory. *Clin. Chim. Acta*, 72: 17-32

Ouyang H., Paschalis E., Mayo W., Boskey A., Mendelsohn R., (2001). *J Bone Miner Res* 16:893–900

Paschalis E., Verdelis K., Doty S., Boskey A., Mendelsohn R., Yanauchi M., (2001). *J Bone Miner Res* 16:1821–1828

Patrick L., (2003). Toxic metals and antioxidants: part II. the role of antioxidants in arsenic and cadmium toxicity. *Altern. Med. Rev.* 8, 106–128

Paulik F., Erö"ss K., Paulik J., Farkas T., Vizkelety T., (1969). Investigation of the composition and crystal structure of bone salt by derivatography and infrared spectrophotometry. *Hoppe Seyler's Z Physiol Chem*.;350:418–26

Prien E.L., (1963). Crystallographic analysis of urinary calculi: a 23-year survey study. *J Urol*; 80: 917–924

Rai A.K., Yueh F.Y., Singh J.P., (2002). High temperature fibre optic laser-induced breakdown spectroscopy sensor for elemental analysis of molten alloy. *Rev Sci Instrum* 73:3589–3599

Rai A.K., Yueh F.Y., Singh J.P., Rai D.K., (2007). Laser induced breakdown spectroscopy for solid and molten materials. In: Singh JP, Thakur SN (eds). *Laser induced breakdown spectroscopy*. Elsevier Science, Netherlands, pp 255–286

Ruegsegger P., Koller B., Muller R., (1996). A microtomographic system for the nondestructive evaluation of bone architecture. *Calcif Tissue Int*, 58(1):24-29

Samek O., Liska M., Kaiser J., Beddows D.C.S., Telle H.H., Kukhlevsky S.V., (2000). Clinical application of laser induced breakdown spectroscopy to the analysis of teeth and dental materials. *Clin Laser Med Surg* 18:281–289

Samek O., Telle H.H., Beddows D.D., (2001). Laser induced breakdown spectroscopy: a tool for real-time, in vitro and in vivo identification of caries teeth. *BMC Oral Health* 1:1–9

Satarug S., Garrett S.H., Sens M.A., Sens D.A., (2010). Cadmium, environmental exposure and health outcomes. *Environ. Health Perspect.* 118, 182–190

Schubert G. (2000). Incidence and analysis of drug-induced urinary stones, In: Rodgers AL, Hibbert BE, Hess B, Khan SR, Preminger GM, eds. *IXth Intern. Symp. Urolithiasis*, publ. University Cape Town, South Africa: 416–418

Schubert G., (1995). Stone analysis of very rare urinary stone components. In: Tiselius HG, ed. *Renal Stones - Aspects on Their Formation, Removal and Prevention*. Stockholm VIth Europ. Symp. Urolithiasis:134–135

Schubert G., Reck G., Jancke H., et al., (2005). Uric acid monohydrate – a new urinary calculus phase. *Urol Res*.;33:238

Shang-Che Lin, Frederick Van Lente, Adam Fadlalla, Walter H. Henricks, (2002). Integration of Text, Image, and Graphic Data From Different Sources in Laboratory Reports: Example of Kidney Stone Reporting System. *Am J Clin Pathol*;118:179-183

Silva S.F.R., Matos D.C., Silva S.L., Daher E.F., Campos H.H., Silva C.A.B., (2010). Chemical and morphological analysis of kidney stones. A double-blind comparative study. *Acta Cir Bras.*; 25(5)

Silva S.F.R., Silva S.L., Daher E.F., Silva Junior G.B., Mota R.M.S., Silva C.A.B., (2009). Determination of urinary stone composition based on stone morphology: a prospective study of 235 consecutive patients in an emerging country. *Clin Chem Lab Med.*;47(5):561-4

Singh V.K., Rai V., Rai A.K., (2009). Variational study of the constituents of cholesterol stones by laser-induced breakdown spectroscopy. *Lasers Med Sci* 24:27–33

Sokol E.V., Nigmatulina E.N., Maksimova N.V., Chiglintsev A.Y., (2003). Spherulites of Calcium Oxalate in Uroliths: Morphology and Formation Conditions. *Chemistry for Sustainable Development* 11: 535–545

Sterling C., (1965). Crystal-structure analysis of weddellite, $CaC_2O_4 * (2+x)H_2O$. *Acta Crystallogr.*, 18, 917- 921

Strates B., Georgacopoulou C., (1969). Derivatographic thermal analysis of renal tract calculi. *Clin Chem.*;15:307–11

Sutor D.J., (1969). Growth studies of calcium oxalate in the presence of various ions and compounds. *Br J Urol* 41:171–178

Sutor D.J., Scheidt   S., (1968). Identification standards for human urinary calculus components, using crystallographic methods. *Br J Urol*; 40: 22–28

Sutor D.J., Wooley S.E., Mackenzie K.R., Wilson R., Scott R., Morgan H.G., (1971). Urinary tract calculi - A comparison of chemical and crystallographic analyses. *Brit J Urol*; 43: 149-153

Takasaki E., (1971). An observation on the analysis of urinary calculi by infrared spectroscopy. *Calc. Tiss. Res.*, 7: 232

Tazzoli V., Domeneghetti C., (1980) The crystal structures of whewellite and weddellite: re-examination and comparison. *American Mineralogist*, Volume 65, pages 27-334

Trinchieri A., Castelnuovo C., Lizzano R., Zanetti G., (2005). Calcium stone disease: a multiform reality. *Urol Res* 33:194–1988

Tsay Y.C., (1961). Application of infrared spectroscopy to analysis of urinary calculi. *J. Urol.*, 86: 838-854

Walther P., Wehrli E., Hermann R., Miller M., (1995). Double layer coating for high-resolution, low-temperature SEM. *J Microsc* 179:229-237

Weissman M., Klein B., Berkowitz J., (1959). Clinical applications of infrared spectroscopy: analysis of renal tract calculi. *Anal. Chem.*, 31: 1334-1338

Welshman S.G., McGeown M.G., (1972). A quantitative investigation of the effects on the growth of calcium oxalate crystals on potential inhibitors. *Br J Urol* 44:677–680

Westbury E.J., (1989). A chemist's view of the history of urinary stone analysis. *Br J Urol* 64:445–450

Zarse C.A., McAteer J.A., Sommer A.J., Kim S.C., Hatt E.K., Lingeman J.E., Evan A.P., Williams J.C. Jr., (2004). Nondestructive analysis of urinary calculi using micro computed Tomography. *BMC Urology*, 4:15-22

Zhou X.S., Shen G.R., Wu J.G., Li W.H., Xu Y.Z., Weng S.F., Soloway R.D., Fu X.B., Tian W., Xu Z., Shen T., Xu G.X., Wentrup Byrne E., (1997). A spectroscopic study of pigment gallstones in China. *Biospectroscopy* 3:371–380

Permissions

The contributors of this book come from diverse backgrounds, making this book a truly international effort. This book will bring forth new frontiers with its revolutionizing research information and detailed analysis of the nascent developments around the world.

We would like to thank Cumhur Aydinalp, for lending his expertise to make the book truly unique. He has played a crucial role in the development of this book. Without his invaluable contribution this book wouldn't have been possible. He has made vital efforts to compile up to date information on the varied aspects of this subject to make this book a valuable addition to the collection of many professionals and students.

This book was conceptualized with the vision of imparting up-to-date information and advanced data in this field. To ensure the same, a matchless editorial board was set up. Every individual on the board went through rigorous rounds of assessment to prove their worth. After which they invested a large part of their time researching and compiling the most relevant data for our readers. Conferences and sessions were held from time to time between the editorial board and the contributing authors to present the data in the most comprehensible form. The editorial team has worked tirelessly to provide valuable and valid information to help people across the globe.

Every chapter published in this book has been scrutinized by our experts. Their significance has been extensively debated. The topics covered herein carry significant findings which will fuel the growth of the discipline. They may even be implemented as practical applications or may be referred to as a beginning point for another development. Chapters in this book were first published by InTech; hereby published with permission under the Creative Commons Attribution License or equivalent.

The editorial board has been involved in producing this book since its inception. They have spent rigorous hours researching and exploring the diverse topics which have resulted in the successful publishing of this book. They have passed on their knowledge of decades through this book. To expedite this challenging task, the publisher supported the team at every step. A small team of assistant editors was also appointed to further simplify the editing procedure and attain best results for the readers.

Our editorial team has been hand-picked from every corner of the world. Their multi-ethnicity adds dynamic inputs to the discussions which result in innovative outcomes. These outcomes are then further discussed with the researchers and contributors who give their valuable feedback and opinion regarding the same. The feedback is then collaborated with the researches and they are edited in a comprehensive manner to aid the understanding of the subject.

Apart from the editorial board, the designing team has also invested a significant amount of their time in understanding the subject and creating the most relevant covers. They scrutinized every image to scout for the most suitable representation of the subject and create an appropriate cover for the book.

The publishing team has been involved in this book since its early stages. They were actively engaged in every process, be it collecting the data, connecting with the contributors or procuring relevant information. The team has been an ardent support to the editorial, designing and production team. Their endless efforts to recruit the best for this project, has resulted in the accomplishment of this book. They are a veteran in the field of academics and their pool of knowledge is as vast as their experience in printing. Their expertise and guidance has proved useful at every step. Their uncompromising quality standards have made this book an exceptional effort. Their encouragement from time to time has been an inspiration for everyone.

The publisher and the editorial board hope that this book will prove to be a valuable piece of knowledge for researchers, students, practitioners and scholars across the globe.

List of Contributors

Cumhur Aydinalp
Uludag University, Bursa, Turkey

B. Maibam
Department of Earth Sciences, Manipur University, Canchipur, India

S. Mitra
Department of Geological Sciences, Jadavpur University, India

J. Gallier and P. Dudoignon
HydrASA Laboratory, ENSIP, Poitiers University, France

J.-M. Hillaireau
INRA Domaine Expérimental, France

C. Castanha
Earth Sciences Division, Lawrence Berkeley National Laboratory, Berkeley, USA

R. Amundson
Division of Ecosystem Sciences, University of California Berkeley, USA

S.E. Trumbore
Max Planck Institute for Biogeochemistry, Jena, Germany

René Duffard
Instituto de Astrofísica de Andalucia – CSIC, Granada, Spain

J.D. Martín-Ramos, A. Cambeses and J.H. Scarrow
University of Granada, Department of Mineralogy and Petrology, Spain

J.L. Díaz-Hernández
IFAPA Camino de Purchil, Departament of Natural Resources, Junta de Andalucía, Spain

A. López-Galindo
Instituto Andaluz de Ciencias de la Tierra (IACT, CSIC-UGR), Spain

Maria Luigia Giannossi and Vito Summa
Laboratory of Environmental and Medical Geology, IMAA-CNR, Italy

Deciphering Mineralogy

T0262191